Mastering Transformers

Build state-of-the-art models from scratch with advanced natural language processing techniques

Savaş Yıldırım

Meysam Asgari-Chenaghlu

BIRMINGHAM—MUMBAI

Mastering Transformers

Copyright © 2021 Packt Publishing

Publishing Product Manager: Aditi Gour

Senior Editor: David Sugarman

Content Development Editor: Nathanya Dias

Technical Editor: Arjun Varma

Copy Editor: Safis Editing

Project Coordinator: Aparna Ravikumar Nair

Proofreader: Safis Editing

Indexer: Tejal Daruwale Soni

Production Designer: Alishon Mendonca

First published: September 2021

Production reference: 1290721

Published by Packt Publishing Ltd.

Livery Place

35 Livery Street

Birmingham

B3 2PB, UK.

ISBN 978-1-80107-765-1

www.packt.com

Contributors

About the authors

Savaş Yıldırım graduated from the Istanbul Technical University Department of Computer Engineering and holds a Ph.D. degree in **Natural Language Processing (NLP)**. Currently, he is an associate professor at the Istanbul Bilgi University, Turkey, and is a visiting researcher at the Ryerson University, Canada. He is a proactive lecturer and researcher with more than 20 years of experience teaching courses on machine learning, deep learning, and NLP. He has significantly contributed to the Turkish NLP community by developing a lot of open source software and resources. He also provides comprehensive consultancy to AI companies on their R&D projects. In his spare time, he writes and directs short films, and enjoys practicing yoga.

First of all, I would like to thank my dear partner, Aylin Oktay, for her continuous support, patience, and encouragement throughout the long process of writing this book. I would also like to thank my colleagues at the Istanbul Bilgi University, Department of Computer Engineering, for their support.

Meysam Asgari-Chenaghlu is an AI manager at Carbon Consulting and is also a Ph.D. candidate at the University of Tabriz. He has been a consultant for Turkey's leading telecommunication and banking companies. He has also worked on various projects, including natural language understanding and semantic search.

First and foremost, I would like to thank my loving and patient wife, Narjes Nikzad-Khasmakhi, for her support and understanding. I would also like to thank my father for his support; may his soul rest in peace. Many thanks to Carbon Consulting and my co-workers.

About the reviewer

Alexander Afanasyev is a software engineer with about 14 years of experience in a variety of different domains and roles. Currently, Alexander is an independent contractor who pursues ideas in the space of computer vision, NLP, and building advanced data collection systems in the cyber threat intelligence domain. Previously, Alexander helped review the *Selenium Testing Cookbook* book by Packt. Outside of daily work, he is an active contributor to Stack Overflow and GitHub.

I would like to thank the authors of this book for their hard work and for providing innovative content; the wonderful team of editors and coordinators with excellent communication skills; and my family, who was and always is supportive of my ideas and my work.

Table of Contents

Section 2: Transformer Models – From Autoencoding to Autoregressive Models

3
Autoencoding Language Models

4
Autoregressive and Other Language Models

Section 3: Advanced Topics

8
Working with Efficient Transformers

9
Cross-Lingual and Multilingual Language Modeling

10
Serving Transformer Models

11
Attention Visualization and Experiment Tracking

Other Books You May Enjoy

Index

Preface

We've seen big changes in **Natural Language Processing** (**NLP**) over the last 20 years. During this time, we have experienced different paradigms and finally entered a new era dominated by the magical transformer architecture. This deep learning architecture has come about by inheriting many approaches. Contextual word embeddings, multi-head self-attention, positional encoding, parallelizable architectures, model compression, transfer learning, and cross-lingual models are among those approaches. Starting with the help of various neural-based NLP approaches, the transformer architecture gradually evolved into an attention-based encoder-decoder architecture and continues to evolve to this day. Now, we are seeing new successful variants of this architecture in the literature. Great models have emerged that use only the encoder part of it, such as BERT, or only the decoder part of it, such as GPT.

Throughout the book, we will touch on these NLP approaches and will be able to work with transformer models easily thanks to the Transformers library from the Hugging Face community. We will provide the solutions step by step to a wide variety of NLP problems, ranging from summarization to question-answering. We will see that we can achieve state-of-the-art results with the help of transformers.

Who this book is for

This book is for deep learning researchers, hands-on NLP practitioners, and machine learning/NLP educators and students who want to start their journey with the transformer architecture. Beginner-level machine learning knowledge and a good command of Python will help you get the most out of this book.

What this book covers

Chapter 1, From Bag-of-Words to the Transformers, provides a brief introduction to the history of NLP, providing a comparison between traditional methods, deep learning models such as CNNs, RNNs, and LSTMs, and transformer models.

Chapter 2, A Hands-On Introduction to the Subject, takes a deeper look at how a transformer model can be used. Tokenizers and models such as BERT will be described with hands-on examples.

Chapter 3, Autoencoding Language Models, is where you will gain knowledge about how to train autoencoding language models on any given language from scratch. This training will include pretraining and the task-specific training of models.

Chapter 4, Autoregressive and Other Language Models, explores the theoretical details of autoregressive language models and teaches you about pretraining them on their own corpus. You will learn how to pretrain any language model such as GPT-2 on their own text and use the model in various tasks such as language generation.

Chapter 5, Fine-Tuning Language Models for Text Classification, is where you will learn how to configure a pre-trained model for text classification and how to fine-tune it for any text classification downstream task, such as sentiment analysis or multi-class classification.

Chapter 6, Fine-Tuning Language Models for Token Classification, teaches you how to fine-tune language models for token classification tasks such as NER, POS tagging, and question-answering.

Chapter 7, Text Representation, is where you will learn about text representation techniques and how to efficiently utilize the transformer architecture, especially for unsupervised tasks such as clustering, semantic search, and topic modeling.

Chapter 8, Working with Efficient Transformers, shows you how to make efficient models out of trained models by using distillation, pruning, and quantization. Then, you will gain knowledge about efficient sparse transformers, such as Linformer and BigBird, and how to work with them.

Chapter 9, Cross-Lingual and Multilingual Language Modeling, is where you will learn about multilingual and cross-lingual language model pretraining and the difference between monolingual and multilingual pretraining. Causal language modeling and translation language modeling are the other topics covered in the chapter.

Chapter 10, Serving Transformer Models, will detail how to serve transformer-based NLP solutions in environments where CPU/GPU is available. Using **TensorFlow Extended** (**TFX**) for machine learning deployment will be described here also.

Chapter 11, Attention Visualization and Experiment Tracking, will cover two different technical concepts: attention visualization and experiment tracking. We will practice them using sophisticated tools such as exBERT and BertViz.

To get the most out of this book

To follow this book, you need to have a basic knowledge of the Python programming language. It is also a required that you know the basics of NLP, deep learning, and how deep neural networks work.

> **Important note**
> All the code in this book has been executed in the Python 3.6 version since some of the libraries in the Python 3.9 version are in development stages.

Software/hardware covered in the book	Operating system requirements
Transformers	Windows, macOS, or Linux
TensorFlow and PyTorch	Windows, macOS, or Linux
Python 3.6x	Windows, macOs or Linux
Jupyter Notebook	Windows, macOs or Linux
Google Colaboratory	Windows, macOs or Linux
Docker	Windows, macOs or Linux
Locust.io	Windows, macOs or Linux
Git	Windows, macOs, or Linux

If you are using the digital version of this book, we advise you to type the code yourself or access the code from the book's GitHub repository (a link is available in the next section). Doing so will help you avoid any potential errors related to the copying and pasting of code.

Download the example code files

You can download the example code files for this book from GitHub at `https://github.com/PacktPublishing/Mastering-Transformers`. If there's an update to the code, it will be updated in the GitHub repository.

We also have other code bundles from our rich catalog of books and videos available at `https://github.com/PacktPublishing/`. Check them out!

Code in Action

The Code in Action videos for this book can be viewed at `https://bit.ly/3i4vFzJ`.

Download the color images

We also provide a PDF file that has color images of the screenshots and diagrams used in this book. You can download it here: `https://static.packt-cdn.com/downloads/9781801077651_ColorImages.pdf`.

Conventions used

There are a number of text conventions used throughout this book.

`Code in text`: Indicates code words in text, database table names, folder names, filenames, file extensions, pathnames, dummy URLs, user input, and Twitter handles. Here is an example: "Sequences that are shorter than `max_sen_len` (maximum sentence length) are padded with a `PAD` value until they are `max_sen_len` in length."

A block of code is set as follows:

```
max_sen_len=max([len(s.split()) for s in sentences])
words = ["PAD"]+ list(set([w for s in sentences for w in
s.split()]))
word2idx= {w:i for i,w in enumerate(words)}
max_words=max(word2idx.values())+1
idx2word= {i:w for i,w in enumerate(words)}
train=[list(map(lambda x:word2idx[x], s.split())) for s in
sentences]
```

When we wish to draw your attention to a particular part of a code block, the relevant lines or items are set in bold:

```
[default]
exten => s,1,Dial(Zap/1|30)
exten => s,2,Voicemail(u100)
exten => s,102,Voicemail(b100)
exten => i,1,Voicemail(s0)
```

Any command-line input or output is written as follows:

```
$ conda activate transformers
$ conda install -c conda-forge tensorflow
```

Bold: Indicates a new term, an important word, or words that you see onscreen. For instance, words in menus or dialog boxes appear in **bold**. Here is an example: "We must now take care of the computational cost of a particular model for a given environment (**Random Access Memory (RAM)**, CPU, and GPU) in terms of memory usage and speed."

> **Tips or important notes**
> Appear like this.

Get in touch

Feedback from our readers is always welcome.

General feedback: If you have questions about any aspect of this book, email us at customercare@packtpub.com and mention the book title in the subject of your message.

Errata: Although we have taken every care to ensure the accuracy of our content, mistakes do happen. If you have found a mistake in this book, we would be grateful if you would report this to us. Please visit www.packtpub.com/support/errata and fill in the form.

Piracy: If you come across any illegal copies of our works in any form on the internet, we would be grateful if you would provide us with the location address or website name. Please contact us at copyright@packt.com with a link to the material.

If you are interested in becoming an author: If there is a topic that you have expertise in and you are interested in either writing or contributing to a book, please visit `authors.packtpub.com`.

Share Your Thoughts

Once you've read *Mastering Transformers*, we'd love to hear your thoughts! Scan the QR code below to go straight to the Amazon review page for this book and share your feedback.

`https://packt.link/r/1-801-07765-7`

Your review is important to us and the tech community and will help us make sure we're delivering excellent quality content.

Section 1: Introduction – Recent Developments in the Field, Installations, and Hello World Applications

In this section, you will learn about all aspects of Transformers at an introductory level. You will write your first `hello-world` program with Transformers by loading community-provided pre-trained language models and running the related code with or without a GPU. Installing and utilizing the `tensorflow`, `pytorch`, `conda`, `transformers`, and `sentenceTransformers` libraries will also be explained in detail in this section.

This section comprises the following chapters:

- *Chapter 1, From Bag-of-Words to the Transformers*
- *Chapter 2, A Hands-On Introduction to the Subject*

1
From Bag-of-Words to the Transformer

In this chapter, we will discuss what has changed in **Natural Language Processing** (**NLP**) over two decades. We experienced different paradigms and finally entered the era of Transformer architectures. All the paradigms help us to gain a better representation of words and documents for problem-solving. Distributional semantics describes the meaning of a word or a document with vectorial representation, looking at distributional evidence in a collection of articles. Vectors are used to solve many problems in both supervised and unsupervised pipelines. For language-generation problems, n-gram language models have been leveraged as a traditional approach for years. However, these traditional approaches have many weaknesses that we will discuss throughout the chapter.

We will further discuss classical **Deep Learning** (**DL**) architectures such as **Recurrent Neural Networks** (**RNNs**), **Feed-Forward Neural Networks** (**FFNNs**), and **Convolutional Neural Networks** (**CNNs**). These have improved the performance of the problems in the field and have overcome the limitation of traditional approaches. However, these models have had their own problems too. Recently, Transformer models have gained immense interest because of their effectiveness in all NLP tasks, from text classification to text generation. However, the main success has been that Transformers effectively improve the performance of multilingual and multi-task NLP problems, as well as monolingual and single tasks. These contributions have made **Transfer Learning** (**TL**) more possible in NLP, which aims to make models reusable for different tasks or different languages.

Starting with the attention mechanism, we will briefly discuss the Transformer architecture and the differences between previous NLP models. In parallel with theoretical discussions, we will show practical examples with the popular NLP framework. For the sake of simplicity, we will choose introductory code examples that are as short as possible.

In this chapter, we will cover the following topics:

- Evolution of NLP toward Transformers
- Understanding distributional semantics
- Leveraging DL
- Overview of the Transformer architecture
- Using TL with Transformers

Technical requirements

We will be using Jupyter Notebook to run our coding exercises that require `python >=3.6.0`, along with the following packages that need to be installed with the `pip install` command:

- `sklearn`
- `nltk==3.5.0`
- `gensim==3.8.3`
- `fasttext`
- `keras>=2.3.0`
- `Transformers >=4.00`

All notebooks with coding exercises are available at the following GitHub link: `https://github.com/PacktPublishing/Advanced-Natural-Language-Processing-with-Transformers/tree/main/CH01`.

Check out the following link to see Code in Action Video: `https://bit.ly/2UFPuVd`

Evolution of NLP toward Transformers

We have seen profound changes in NLP over the last 20 years. During this period, we experienced different paradigms and finally entered a new era dominated mostly by magical *Transformer* architecture. This architecture did not come out of nowhere. Starting with the help of various neural-based NLP approaches, it gradually evolved to an attention-based encoder-decoder type architecture and still keeps evolving. The architecture and its variants have been successful thanks to the following developments in the last decade:

- Contextual word embeddings

- Better subword tokenization algorithms for handling unseen words or rare words

- Injecting additional memory tokens into sentences, such as `Paragraph ID` in `Doc2vec` or a **Classification (CLS)** token in **Bidirectional Encoder Representations from Transformers** (**BERT**)

- Attention mechanisms, which overcome the problem of forcing input sentences to encode all information into one context vector

- Multi-head self-attention

- Positional encoding to case word order

- Parallelizable architectures that make for faster training and fine-tuning

- Model compression (distillation, quantization, and so on)

- TL (cross-lingual, multitask learning)

For many years, we used traditional NLP approaches such as *n-gram language models*, *TF-IDF-based information retrieval models*, and *one-hot encoded document-term matrices*. All these approaches have contributed a lot to the solution of many NLP problems such as *sequence classification*, *language generation*, *language understanding*, and so forth. On the other hand, these traditional NLP methods have their own weaknesses—for instance, falling short in solving the problems of sparsity, unseen words representation, tracking long-term dependencies, and others. In order to cope with these weaknesses, we developed DL-based approaches such as the following:

- RNNs

- CNNs

- FFNNs

- Several variants of RNNs, CNNs, and FFNNs

In 2013, as a two-layer FFNN word-encoder model, `Word2vec`, sorted out the dimensionality problem by producing short and dense representations of the words, called **word embeddings**. This early model managed to produce fast and efficient static word embeddings. It transformed unsupervised textual data into supervised data (*self-supervised learning*) by either predicting the target word using context or predicting neighbor words based on a sliding window. **GloVe**, another widely used and popular model, argued that count-based models can be better than neural models. It leverages both global and local statistics of a corpus to learn embeddings based on word-word co-occurrence statistics. It performed well on some syntactic and semantic tasks, as shown in the following screenshot. The screenshot tells us that the embeddings offsets between the terms help to apply vector-oriented reasoning. We can learn the generalization of gender relations, which is a semantic relation from the offset between *man* and *woman* (*man-> woman*). Then, we can arithmetically estimate the vector of *actress* by adding the vector of the term *actor* and the offset calculated before. Likewise, we can learn syntactic relations such as word plural forms. For instance, if the vectors of **Actor**, **Actors**, and **Actress** are given, we can estimate the vector of **Actresses**:

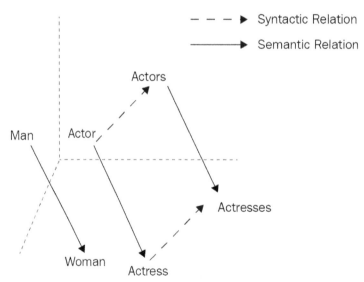

Figure 1.1 – Word embeddings offset for relation extraction

The recurrent and convolutional architectures such as RNN, **Long Short-Term Memory (LSTM)**, and CNN started to be used as encoders and decoders in **sequence-to-sequence (seq2seq)** problems. The main challenge with these early models was polysemous words. The senses of the words are ignored since a single fixed representation is assigned to each word, which is especially a severe problem for polysemous words and sentence semantics.

The further pioneer neural network models such as **Universal Language Model Fine-tuning (ULMFit)** and **Embeddings from Language Models (ELMo)** managed to encode the sentence-level information and finally alleviate polysemy problems, unlike with static word embeddings. These two important approaches were based on LSTM networks. They also introduced the concept of pre-training and fine-tuning. They help us to apply TL, employing the pre-trained models trained on a general task with huge textual datasets. Then, we can easily perform fine-tuning by resuming training of the pre-trained network on a target task with supervision. The representations differ from traditional word embeddings such that each word representation is a function of the entire input sentence. The modern Transformer architecture took advantage of this idea.

In the meantime, the idea of an attention mechanism made a strong impression in the NLP field and achieved significant success, especially in seq2seq problems. Earlier methods would pass the last state (known as a **context vector** or **thought vector**) obtained from the entire input sequence to the output sequence without linking or elimination. The attention mechanism was able to build a more sophisticated model by linking the tokens determined from the input sequence to the particular tokens in the output sequence. For instance, suppose you have a keyword phrase `Government of Canada` in the input sentence for an English to Turkish translation task. In the output sentence, the `Kanada Hükümeti` token makes strong connections with the input phrase and establishes a weaker connection with the remaining words in the input, as illustrated in the following screenshot:

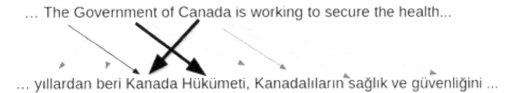

Figure 1.2 – Sketchy visualization of an attention mechanism

So, this mechanism makes models more successful in seq2seq problems such as translation, question answering, and text summarization.

In 2017, the Transformer-based encoder-decoder model was proposed and found to be successful. The design is based on an FFNN by discarding RNN recurrency and using only attention mechanisms (*Vaswani et al., All you need is attention, 2017*). The Transformer-based models have so far overcome many difficulties that other approaches faced and have become a new paradigm. Throughout this book, you will be exploring and understanding how the Transformer-based models work.

Understanding distributional semantics

Distributional semantics describes the meaning of a word with a vectorial representation, preferably looking at its distributional evidence rather than looking at its predefined dictionary definitions. The theory suggests that words co-occurring together in a similar environment tend to share similar meanings. This was first formulated by the scholar Harris (*Distributional Structure Word, 1954*). For example, similar words such as *dog* and *cat* mostly co-occur in the same context. One of the advantages of a distributional approach is to help the researchers to understand and monitor the semantic evolution of words across time and domains, also known as the **lexical semantic change problem**.

Traditional approaches have applied **Bag-of-Words** (**BoW**) and n-gram language models to build the representation of words and sentences for many years. In a BoW approach, words and documents are represented with a one-hot encoding as a sparse way of representation, also known as the **Vector Space Model** (**VSM**).

Text classification, word similarity, semantic relation extraction, word-sense disambiguation, and many other NLP problems have been solved by these one-hot encoding techniques for years. On the other hand, n-gram language models assign probabilities to sequences of words so that we can either compute the probability that a sequence belongs to a corpus or generate a random sequence based on a given corpus.

BoW implementation

A BoW is a representation technique for documents by counting the words in them. The main data structure of the technique is a document-term matrix. Let's see a simple implementation of BoW with Python. The following piece of code illustrates how to build a document-term matrix with the Python `sklearn` library for a toy corpus of three sentences:

```
from sklearn.feature_extraction.text import TfidfVectorizer
import numpy as np
import pandas as pd
toy_corpus= ["the fat cat sat on the mat",
```

```
                "the big cat slept",
                "the dog chased a cat"]
vectorizer=TfidfVectorizer()
corpus_tfidf=vectorizer.fit_transform(toy_corpus)
print(f"The vocabulary size is \
                    {len(vectorizer.vocabulary_.keys())} ")
print(f"The document-term matrix shape is\
                    {corpus_tfidf.shape}")
df=pd.DataFrame(np.round(corpus_tfidf.toarray(),2))
df.columns=vectorizer.get_feature_names()
```

The output of the code is a document-term matrix, as shown in the following screenshot. The size is (3 x 10), but in a realistic scenario the matrix size can grow to much larger numbers such as 10K x 10M:

```
The vocabulary size is 10
The document-term matrix shape is (3, 10)
      big   cat  chased  dog  fat   mat    on   sat  slept   the

 0   0.00  0.25   0.00  0.00  0.42  0.42  0.42  0.42   0.00  0.49

 1   0.61  0.36   0.00  0.00  0.00  0.00  0.00  0.00   0.61  0.36

 2   0.00  0.36   0.61  0.61  0.00  0.00  0.00  0.00   0.00  0.36
```

Figure 1.3 – Document-term matrix

The table indicates a count-based mathematical matrix where the cell values are transformed by a **Term Frequency-Inverse Document Frequency (TF-IDF)** weighting schema. This approach does not care about the position of words. Since the word order strongly determines the meaning, ignoring it leads to a loss of meaning. This is a common problem in a BoW method, which is finally solved by a recursion mechanism in RNN and positional encoding in Transformers.

Each column in the matrix stands for the vector of a word in the vocabulary, and each row stands for the vector of a document. Semantic similarity metrics can be applied to compute the similarity or dissimilarity of the words as well as documents. Most of the time, we use bigrams such as cat_sat and the_street to enrich the document representation. For instance, as the parameter ngram_range=(1,2) is passed to TfidfVectorizer, it builds a vector space containing both unigrams (big, cat, dog) and bigrams (big_cat, big_dog). Thus, such models are also called **bag-of-n-grams**, which is a natural extension of **BoW**.

If a word is commonly used in each document, it can be considered to be high-frequency, such as *and the*. Conversely, some words hardly appear in documents, called low-frequency (or rare) words. As high-frequency and low-frequency words may prevent the model from working properly, TF-IDF, which is one of the most important and well-known weighting mechanisms, is used here as a solution.

Inverse Document Frequency (IDF) is a statistical weight to measure the importance of a word in a document—for example, while the word the has no discriminative power, chased can be highly informative and give clues about the subject of the text. This is because high-frequency words (stopwords, functional words) have little discriminating power in understanding the documents.

The discriminativeness of the terms also depends on the domain—for instance, a list of DL articles is most likely to have the word network in almost every document. IDF can scale down the weights of all terms by using their **Document Frequency (DF)**, where the DF of a word is computed by the number of documents in which a term appears. **Term Frequency (TF)** is the raw count of a term (word) in a document.

Some of the advantages and disadvantages of a TF-IDF based BoW model are listed as follows:

Advantages	Disadvantages
• Easy to implement • Human-interpretable results • Domain adaptation	• Dimensionality curse. • No solution for unseen words. • Hardly capture semantic relations, such as is-a, has-a, synonym. • Word order information is ignored. • Slow for large vocabularies.

Table 1 – Advantages and disadvantages of a TF-IDF BoW model

Overcoming the dimensionality problem

To overcome the dimensionality problem of the BoW model, **Latent Semantic Analysis (LSA)** is widely used for capturing semantics in a low-dimensional space. It is a linear method that captures pairwise correlations between terms. LSA-based probabilistic methods can be still considered as a single layer of hidden topic variables. However, current DL models include multiple hidden layers, with billions of parameters. In addition to that, Transformer-based models showed that they can discover latent representations much better than such traditional models.

For the **Natural Language Understanding** (**NLU**) tasks, the traditional pipeline starts with some preparation steps, such as *tokenization*, *stemming*, *noun phrase detection*, *chunking*, *stop-word elimination*, and much more. Afterward, a document-term matrix is constructed with any weighting schema, where TF-IDF is the most popular one. Finally, the matrix is served as a tabulated input for **Machine Learning** (**ML**) pipelines, sentiment analysis, document similarity, document clustering, or measuring the relevancy score between a query and a document. Likewise, terms are represented as a tabular matrix and can be input for a token classification problem where we can apply named-entity recognition, semantic relation extractions, and so on.

The classification phase includes a straightforward implementation of supervised ML algorithms such as **Support Vector Machine** (**SVM**), Random forest, logistic, naive bayes, and Multiple Learners (Boosting or Bagging). Practically, the implementation of such a pipeline can simply be coded as follows:

```
from sklearn.pipeline import make_pipeline
from sklearn.svm import SVC
labels= [0,1,0]
clf = SVC()
clf.fit(df.to_numpy(), labels)
```

As seen in the preceding code, we can apply fit operations easily thanks to the `sklearn` **Application Programming Interface** (**API**). In order to apply the learned model to train data, the following code is executed:

```
clf.predict(df.to_numpy())
Output: array([0, 1, 0])
```

Let's move on to the next section!

Language modeling and generation

For language-generation problems, the traditional approaches are based on leveraging n-gram language models. This is also called a **Markov process**, which is a stochastic model in which each word (event) depends on a subset of previous words—*unigram*, *bigram*, or *n-gram*, outlined as follows:

- **Unigram** (all words are independent and no chain): This estimates the probability of word in a vocabulary simply computed by the frequency of it to the total word count.

- **Bigram** (First-order Markov process): This estimates the $P(word_i | wordi_{-1})$. probability of *wordi* depending on *wordi-1*, which is simply computed by the ratio of $P(word_i, wordi_{-1})$ to $P(wordi_{-1})$.

- **Ngram** (N-order Markov process): This estimates $P(wordi | word0, ..., wordi-1)$.

Let's give a simple language model implementation with the **Natural Language Toolkit (NLTK)** library. In the following implementation, we train a **Maximum Likelihood Estimator (MLE)** with order $n=2$. We can select any n-gram order such as $n=1$ for unigrams, $n=2$ for bigrams, $n=3$ for trigrams, and so forth:

```
import nltk
from nltk.corpus import gutenberg
from nltk.lm import MLE
from nltk.lm.preprocessing import padded_everygram_pipeline
nltk.download('gutenberg')
nltk.download('punkt')
macbeth = gutenberg.sents('shakespeare-macbeth.txt')
model, vocab = padded_everygram_pipeline(2, macbeth)
lm=MLE(2)
lm.fit(model,vocab)
print(list(lm.vocab)[:10])
print(f"The number of words is {len(lm.vocab)}")
```

The nltk package first downloads the gutenberg corpus, which includes some texts from the *Project Gutenberg* electronic text archive, hosted at https://www.gutenberg.org. It also downloads the punkt tokenizer tool for the punctuation process. This tokenizer divides a raw text into a list of sentences by using an unsupervised algorithm. The nltk package already includes a pre-trained English punkt tokenizer model for abbreviation words and collocations. It can be trained on a list of texts in any language before use. In the further chapters, we will discuss how to train different and more efficient tokenizers for Transformer models as well. The following code produces what the language model learned so far:

```
print(f"The frequency of the term 'Macbeth' is {lm.
counts['Macbeth']}")
print(f"The language model probability score of 'Macbeth' is
{lm.score('Macbeth')}")
print(f"The number of times 'Macbeth' follows 'Enter' is {lm.
counts[['Enter']]['Macbeth']} ")
print(f"P(Macbeth | Enter) is {lm.score('Macbeth',
```

```
['Enter'])}")
print(f"P(shaking | for) is {lm.score('shaking', ['for'])}")
```

This is the output:

```
The frequency of the term 'Macbeth' is 61
The language model probability score of 'Macbeth' is 0.00226
The number of times 'Macbeth' follows 'Enter' is 15
P(Macbeth | Enter) is 0.1875
P(shaking | for) is 0.0121
```

The n-gram language model keeps *n-gram* counts and computes the conditional probability for sentence generation. `lm=MLE(2)` stands for MLE, which yields the maximum probable sentence from each token probability. The following code produces a random sentence of 10 words with the `<s>` starting condition given:

```
lm.generate(10, text_seed=['<s>'], random_seed=42)
```

The output is shown in the following snippet:

```
['My', 'Bosome', 'franchis', "'", 's', 'of', 'time', ',', 'We',
'are']
```

We can give a specific starting condition through the `text_seed` parameter, which makes the generation be conditioned on the preceding context. In our preceding example, the preceding context is `<s>`, which is a special token indicating the beginning of a sentence.

So far, we have discussed paradigms underlying traditional NLP models and provided very simple implementations with popular frameworks. We are now moving to the DL section to discuss how neural language models shaped the field of NLP and how neural models overcome the traditional model limitations.

Leveraging DL

NLP is one of the areas where DL architectures have been widely and successfully used. For decades, we have witnessed successful architectures, especially in word and sentence representation. In this section, we will share the story of these different approaches with commonly used frameworks.

Learning word embeddings

Neural network-based language models effectively solved feature representation and language modeling problems since it became possible to train more complex neural architecture on much larger datasets to build short and dense representations. In 2013, the **Word2vec model**, which is a popular word-embedding technique, used a simple and effective architecture to learn a high quality of continuous word representations. It outperformed other models for a variety of syntactic and semantic language tasks such as *sentiment analysis, paraphrase detection, relation extraction*, and so forth. The other key factor in the popularity of the model is its much *lower computational complexity*. It maximizes the probability of the current word given any surrounding context words, or vice versa.

The following piece of code illustrates how to train word vectors for the sentences of the play *Macbeth*:

```
from gensim.models import Word2vec
model = Word2vec(sentences=macbeth, size=100, window= 4, min_
count=10, workers=4, iter=10)
```

The code trains the word embeddings with a vector size of 100 by a sliding 5-length context window. To visualize the words embeddings, we need to reduce the dimension to 3 by applying **Principal Component Analysis (PCA)** as shown in the following code snippet:

```
import matplotlib.pyplot as plt
from sklearn.decomposition import PCA
import random
np.random.seed(42)
words=list([e for e in model.wv.vocab if len(e)>4])
random.shuffle(words)
words3d = PCA(n_components=3,random_state=42).fit_
transform(model.wv[words[:100]])
def plotWords3D(vecs, words, title):
    ...

plotWords3D(words3d, words, "Visualizing Word2vec Word
Embeddings using PCA")
```

This is the output:

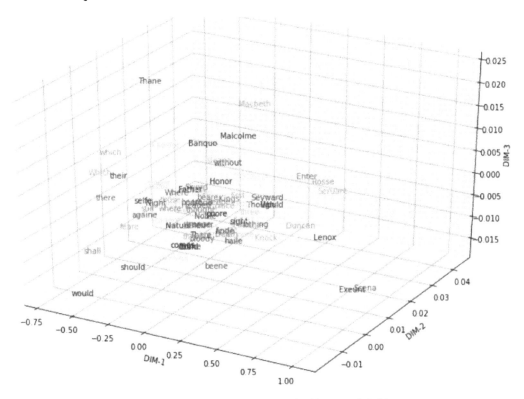

Figure 1.4 – Visualizing word embeddings with PCA

As the plot shows, the main characters of Shakespeare's play—**Macbeth**, **Malcolm**, **Banquo**, **Macduff**, and others—are mapped close to each other. Likewise, auxiliary verbs **shall**, **should**, and **would** appear close to each other at the left-bottom of *Figure 1.4*. We can also capture an analogy such as *man-woman= uncle-aunt* by using an embedding offset. For more interesting visual examples on this topic, please check the following project: `https://projector.tensorflow.org/`.

The Word2vec-like models learn word embeddings by employing a prediction-based neural architecture. They employ gradient descent on some objective functions and nearby word predictions. While traditional approaches apply a count-based method, neural models design a prediction-based architecture for distributional semantics. *Are count-based methods or prediction-based methods the best for distributional word representations?* The GloVe approach addressed this problem and argued that these two approaches are not dramatically different. Jeffrey Penington et al. even supported the idea that the count-based methods could be more successful by capturing global statistics. They stated that GloVe outperformed other neural network language models on word analogy, word similarity, and **Named Entity Recognition** (**NER**) tasks.

These two paradigms, however, did not provide a helpful solution for unseen words and word-sense problems. They do not exploit subword information, and therefore cannot learn the embeddings of rare and unseen words.

FastText, another widely used model, proposed a new enriched approach using subword information, where each word is represented as a bag of character n-grams. The model sets a constant vector to each character n-gram and represents words as the sum of their sub-vectors, which is an idea that was first introduced by Hinrich Schütze (*Word Space, 1993*). The model can compute word representations even for unseen words and learn the internal structure of words such as suffixes/affixes, which is especially important with morphologically rich languages such as Finnish, Hungarian, Turkish, Mongolian, Korean, Japanese, Indonesian, and so forth. Currently, modern Transformer architectures use a variety of subword tokenization methods such as **WordPiece**, **SentencePiece**, or **Byte-Pair Encoding** (**BPE**).

A brief overview of RNNs

RNN models can learn each token representation by rolling up the information of other tokens at an earlier timestep and learn sentence representation at the last timestep. This mechanism has been found beneficial in many ways, outlined as follows:

- Firstly, RNN can be redesigned in a one-to-many model for language generation or music generation.

- Secondly, many-to-one models can be used for text classification or sentiment analysis.

- And lastly, many-to-many models are used for NER problems. The second use of many-to-many models is to solve encoder-decoder problems such as *machine translation*, *question answering*, and *text summarization*.

As with other neural network models, RNN models take tokens produced by a tokenization algorithm that breaks down the entire raw text into atomic units also called tokens. Further, it associates the token units with numeric vectors—token embeddings—which are learned during the training. As an alternative, we can assign the embedded learning task to the well-known word-embedding algorithms such as Word2vec or FastText in advance.

Here is a simple example of an RNN architecture for the sentence `The cat is sad.`, where x_0 is the vector embeddings of `the`, x_1 is the vector embeddings of `cat`, and so forth. *Figure 1.5* illustrates an RNN being unfolded into a full **Deep Neural Network (DNN)**.

Unfolding means that we associate a layer to each word. For the `The cat is sad.` sequence, we take care of a sequence of five words. The hidden state in each layer acts as the memory of the network. It encodes information about what happened in all previous timesteps and in the current timestep. This is represented in the following diagram:

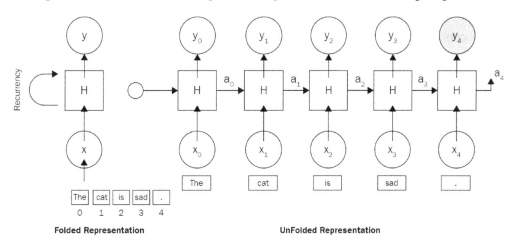

Figure 1.5 – An RNN architecture

The following are some advantages of an RNN architecture:

- **Variable-length input**: The capacity to work on variable-length input, no matter the size of the sentence being input. We can feed the network with sentences of 3 or 300 words without changing the parameter.

- **Caring about word order**: It processes the sequence word by word in order, caring about the word position.

- **Suitable for working in various modes (many-to-many, one-to-many)**: We can train a machine translation model or sentiment analysis using the same recurrency paradigm. Both architectures would be based on an RNN.

The disadvantages of an RNN architecture are listed here:

- **Long-term dependency problem**: When we process a very long document and try to link the terms that are far from each other, we need to care about and encode all irrelevant other terms between these terms.

- **Prone to exploding or vanishing gradient problems**: When working on long documents, updating the weights of the very first words is a big deal, which makes a model untrainable due to a vanishing gradient problem.

- **Hard to apply parallelizable training**: Parallelization breaks the main problem down into a smaller problem and executes the solutions at the same time, but RNN follows a classic sequential approach. Each layer strongly depends on previous layers, which makes parallelization impossible.

- **The computation is slow as the sequence is long**: An RNN could be very efficient for short text problems. It processes longer documents very slowly, besides the long-term dependency problem.

Although an RNN can theoretically attend the information at many timesteps before, in the real world, problems such as long documents and long-term dependencies are impossible to discover. Long sequences are represented within many deep layers. These problems have been addressed by many studies, some of which are outlined here:

- *Hochreiter and Schmidhuber. Long Short-term Memory. 1997.*

- *Bengio et al. Learning long-term dependencies with gradient descent is difficult. 1993.*

- *K. Cho et al. Learning phrase representations using RNN encoder-decoder for statistical machine translation. 2014.*

LSTMs and gated recurrent units

LSTM (*Schmidhuber, 1997*) and **Gated Recurrent Units (GRUs)** (*Cho, 2014*) are new variants of RNNs, have solved long-term dependency problems, and have attracted great attention. LSTMs were particularly developed to cope with the long-term dependency problem. The advantage of an LSTM model is that it uses the additional cell state, which is a horizontal sequence line on the top of the LSTM unit. This cell state is controlled by special purpose gates for forget, insert, or update operations. The complex unit of an LSTM architecture is depicted in the following diagram:

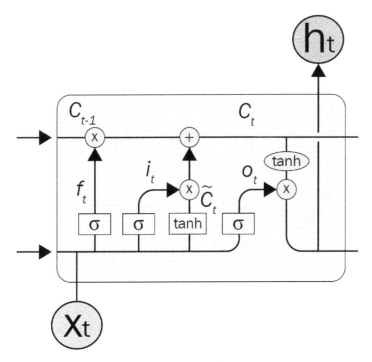

Figure 1.6 – An LSTM unit

It is able to decide the following:

- What kind of information we will store in the cell state

- Which information will be forgotten or deleted

In the original RNN, in order to learn the state of I tokens, it recurrently processes the entire state of previous tokens between $timestep_0$ and $timestepi_{-1}$. Carrying entire information from earlier timesteps leads to vanishing gradient problems, which makes the model untrainable. The gate mechanism in LSTM allows the architecture to skip some unrelated tokens at a certain timestep or remember long-range states in order to learn the current token state.

A GRU is similar to an LSTM in many ways, the main difference being that a GRU does not use the cell state. Rather, the architecture is simplified by transferring the functionality of the cell state to the hidden state, and it only includes two gates: an *update gate* and a *reset gate*. The update gate determines how much information from the previous and current timesteps will be pushed forward. This feature helps the model keep relevant information from the past, which minimizes the risk of a vanishing gradient problem as well. The reset gate detects the irrelevant data and makes the model forget it.

A gentle implementation of LSTM with Keras

We need to download the **Stanford Sentiment Treebank** (**SST-2**) sentiment dataset from the **General Language Understanding Evaluation** (**GLUE**) benchmark. We can do this by running the following code:

```
$ wget https://dl.fbaipublicfiles.com/glue/data/SST-2.zip
$ unzip SST-2.zip
```

> **Important note**
>
> **SST-2**: This is a fully labeled parse tree that allows for complete sentiment analysis in English. The corpus originally consists of about 12K single sentences extracted from movie reviews. It was parsed with the Stanford parser and includes over 200K unique phrases, each annotated by three human judges. For more information, see *Socher et al., Parsing With Compositional Vector Grammars, EMNLP. 2013* (https://nlp.stanford.edu/sentiment).

After downloading the data, let's read it as a pandas object, as follows:

```
import tensorflow as tf
import pandas as pd
df=pd.read_csv('SST-2/train.tsv',sep="\t")
sentences=df.sentence
labels=df.label
```

We need to set maximum sentence length, build vocabulary and dictionaries (word2idx, idx2words), and finally represent each sentence as a list of indexes rather than strings. We can do this by running the following code:

```
max_sen_len=max([len(s.split()) for s in sentences])
words = ["PAD"]+\
        list(set([w for s in sentences for w in s.split()]))
word2idx= {w:i for i,w in enumerate(words)}
max_words=max(word2idx.values())+1
idx2word= {i:w for i,w in enumerate(words)}
train=[list(map(lambda x:word2idx[x], s.split()))\
                            for s in sentences]
```

Sequences that are shorter than `max_sen_len` (maximum sentence length) are padded with a `PAD` value until they are `max_sen_len` in length. On the other hand, longer sequences are truncated so that they fit `max_sen_len`. Here is the implementation:

```
from keras import preprocessing
train_pad = preprocessing.sequence.pad_sequences(train,
                                    maxlen=max_sen_len)
print('Train shape:', train_pad.shape)
Output: Train shape: (67349, 52)
```

We are ready to design and train an LSTM model, as follows:

```
from keras.layers import LSTM, Embedding, Dense
from keras.models import Sequential
model = Sequential()
model.add(Embedding(max_words, 32))
model.add(LSTM(32))
model.add(Dense(1, activation='sigmoid'))
model.compile(optimizer='rmsprop',loss='binary_crossentropy',
metrics=['acc'])
history = model.fit(train_pad,labels, epochs=30, batch_size=32,
validation_split=0.2)
```

The model will be trained for 30 epochs. In order to plot what the LSTM model has learned so far, we can execute the following code:

```
import matplotlib.pyplot as plt
def plot_graphs(history, string):
    ...
plot_graphs(history, 'acc')
plot_graphs(history, 'loss')
```

The code produces the following plot, which shows us the training and validation performance of the LSTM-based text classification:

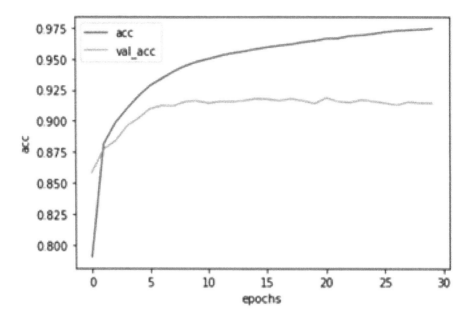

Figure 1.7 – The classification performance of the LSTM network

As we mentioned before, the main problem of an RNN-based encoder-decoder model is that it produces a single fixed representation for a sequence. However, the attention mechanism allowed the RNN to focus on certain parts of the input tokens as it maps them to a certain part of the output tokens. This attention mechanism has been found to be useful and has become one of the underlying ideas of the Transformer architecture. We will discuss how the Transformer architecture takes advantage of attention in the next part and throughout the entire book.

A brief overview of CNNs

CNNs, after their success in computer vision, were ported to NLP in terms of modeling sentences or tasks such as semantic text classification. A CNN is composed of convolution layers followed by a dense neural network in many practices. A convolution layer performs over the data in order to extract useful features. As with any DL model, a convolution layer plays the feature extraction role to automate feature extraction. This feature layer, in the case of NLP, is fed by an embedding layer that takes sentences as an input in a one-hot vectorized format. The one-hot vectors are generated by a `token-id` for each word forming a sentence. The left part of the following screenshot shows a one-hot representation of a sentence:

	1	2	3	4	5
I	1	0	0	0	0
saw	0	1	0	0	0
a	0	0	1	0	0
cat	0	0	0	1	0
.	0	0	0	0	1

Figure 1.8 – One-hot vectors

Each token, represented by a one-hot vector, is fed to the embedding layer. The embedding layer can be initialized by random values or by using pre-trained word vectors such as GloVe, Word2vec, or FastText. A sentence will then be transformed into a dense matrix in the shape of NxE (where **N** is the number of tokens in a sentence and **E** is the embedding size). The following screenshot illustrates how a 1D CNN processes that dense matrix:

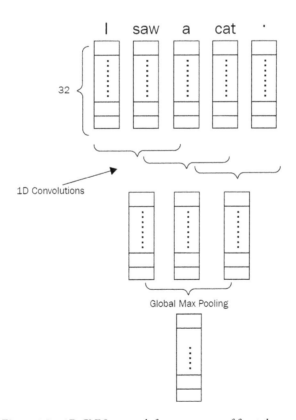

Figure 1.9 – 1D CNN network for a sentence of five tokens

Convolution will take place on top of this operation with different layers and kernels. Hyperparameters for the convolution layer are the kernel size and the number of kernels. It is also good to note that 1D convolution is applied here and the reason for that is token embeddings cannot be seen as partial, and we want to apply kernels capable of seeing multiple tokens in a sequential order together. You can see it as something like an n-gram with a specified window. Using shallow TL combined with CNN models is also another good capability of such models. As shown in the following screenshot, we can also propagate the networks with a combination of many representations of tokens, as proposed in the 2014 study by Yoon Kim, *Convolutional Neural Networks for Sentence Classification*:

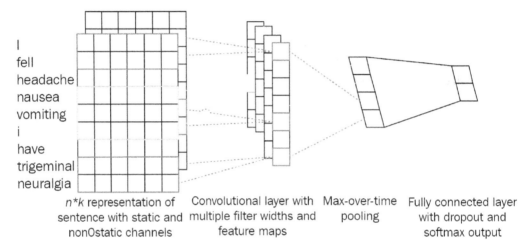

Figure 1.10 – Combination of many representations in a CNN

For example, we can use three embedding layers instead of one and concatenate them for each token. Given this setup, a token such as **fell** will have a vector size of 3x128 if the embedding size is 128 for all three different embeddings. These embeddings can be initialized with pre-trained vectors from Word2vec, GloVe, and FastText. The convolution operation at each step will see N words with their respective three vectors (N is the convolution filter size). The type of convolution that is used here is a 1D convolution. The dimension here denotes possible movements when doing the operation. For example, a 2D convolution will move along two axes, while a 1D convolution just moves along one axis. The following screenshot shows the differences between them:

Conv Direction	Input	Filter	Output
1-direction	3-dim	3-dim	2-dim
2-direction	4-dim	4-dim	3-dim
3-direction	5-dim	5-dim	4-dim

Figure 1.11 – Convolutional directions

The following code snippet is a 1D CNN implementation processing the same data used in an LSTM pipeline. It includes a composition of `Conv1D` and `MaxPooling` layers, followed by `GlobalMaxPooling` layers. We can extend the pipeline by tweaking the parameters and adding more layers to optimize the model:

```
from keras import layers
model = Sequential()
model.add(layers.Embedding(max_words, 32, input_length=max_sen_
len))
model.add(layers.Conv1D(32, 8, activation='relu'))
model.add(layers.MaxPooling1D(4))
model.add(layers.Conv1D(32, 3, activation='relu'))
model.add(layers.GlobalMaxPooling1D())
model.add(layers.Dense(1, activation= 'sigmoid')
model.compile(loss='binary_crossentropy', metrics=['acc'])
history = model.fit(train_pad,labels, epochs=15, batch_size=32,
validation_split=0.2)
```

It turns out that the CNN model showed comparable performance with its LSTM counterpart. Although CNNs have become a standard in image processing, we have seen many successful applications of CNNs for NLP. While an LSTM model is trained to recognize patterns across time, a CNN model recognizes patterns across space.

Overview of the Transformer architecture

Transformer models have received immense interest because of their effectiveness in an enormous range of NLP problems, from text classification to text generation. The attention mechanism is an important part of these models and plays a very crucial role. Before Transformer models, the attention mechanism was proposed as a helper for improving conventional DL models such as RNNs. To have a good understanding of Transformers and their impact on the NLP, we will first study the attention mechanism.

Attention mechanism

One of the first variations of the attention mechanism was proposed by *Bahdanau et al. (2015)*. This mechanism is based on the fact that RNN-based models such as GRUs or LSTMs have an information bottleneck on tasks such as **Neural Machine Translation** (**NMT**). These encoder-decoder-based models get the input in the form of a `token-id` and process it in a recurrent fashion (encoder). Afterward, the processed intermediate representation is fed into another recurrent unit (decoder) to extract the results. This avalanche-like information is like a rolling ball that consumes all the information, and rolling it out is hard for the decoder part because the decoder part does not see all the dependencies and only gets the intermediate representation (context vector) as an input.

To align this mechanism, Bahdanau proposed an attention mechanism to use weights on intermediate hidden values. These weights align the amount of attention a model must pay to input in each decoding step. Such wonderful guidance assists models in specific tasks such as NMT, which is a many-to-many task. A diagram of a typical attention mechanism is provided here:

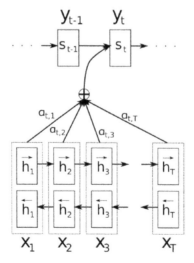

Figure 1.12 – Attention mechanism

Different attention mechanisms have been proposed with different improvements. *Additive, multiplicative, general,* and *dot-product* attention appear within the family of these mechanisms. The latter, which is a modified version with a scaling parameter, is noted as scaled dot-product attention. This specific attention type is the foundation of Transformers models and is called a **multi-head attention mechanism**. Additive attention is also what was introduced earlier as a notable change in NMT tasks. You can see an overview of the different types of attention mechanisms here:

Name	Attention score function	Citation
Content-based attention	$\text{score}(\boldsymbol{s}_t, \boldsymbol{h}_i) = \text{cosine}[\boldsymbol{s}_t, \boldsymbol{h}_i]$	Graves2014
Additive	$\text{score}(\boldsymbol{s}_t, \boldsymbol{h}_i) = \boldsymbol{v}_a^\top \tanh(\mathbf{W}_a[\boldsymbol{s}_t; \boldsymbol{h}_i])$	Bahdanau2015
Location-base	$\alpha_{t,i} = \text{softmax}(\mathbf{W}_a \boldsymbol{s}_t)$	Loung2015
General	$\text{score}(\boldsymbol{s}_t, \boldsymbol{h}_i) = \boldsymbol{s}_t^\top \mathbf{W}_a \boldsymbol{h}_i$	Loung2015
Dot-product	$\text{score}(\boldsymbol{s}_t, \boldsymbol{h}_i) = \boldsymbol{s}_t^\top \boldsymbol{h}_i$	Loung2015
Scaled dot-product	$\text{score}(\boldsymbol{s}_t, \boldsymbol{h}_i) = \dfrac{\boldsymbol{s}_t^\top \boldsymbol{h}_i}{\sqrt{n}}$	Vaswani2017

Table 2 – Types of attention mechanisms (Image inspired from https://lilianweng.github.io/lil-log/2018/06/24/attention-attention.html)

Since attention mechanisms are not specific to NLP, they are also used in different use cases in various fields, from computer vision to speech recognition. The following screenshot shows a visualization of a multimodal approach trained for neural image captioning (*K Xu et al., Show, attend and tell: Neural image caption generation with visual attention, 2015*):

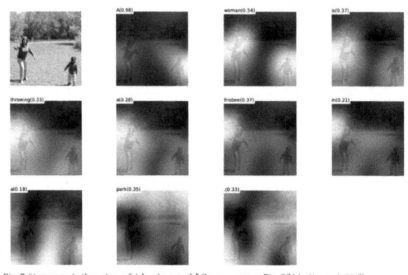

Fig. 7. "A woman is throwing a frisbee in a park." (Image source: Fig. 6(b) in Xu et al. 2015)

Figure 1.13 – Attention mechanism in computer vision

The multi-head attention mechanism that is shown in the following diagram is an essential part of the Transformer architecture:

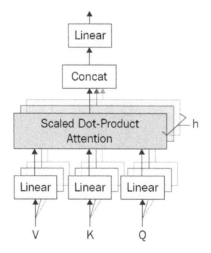

Figure 1.14 – Multi-head attention mechanism

Next, let's understand multi-head attention mechanisms.

Multi-head attention mechanisms

Before jumping into scaled dot-product attention mechanisms, it's better to get a good understanding of self-attention. **Self-attention**, as shown in *Figure 1.15*, is a basic form of a scaled self-attention mechanism. This mechanism uses an input matrix shown as X and produces an attention score between various items in X. We see X as a 3x4 matrix where 3 represents the number of tokens and 4 presents the embedding size. Q from *Figure 1.15* is also known as the **query**, K is known as the **key**, and V is noted as the **value**. Three types of matrices shown as *theta*, *phi*, and g are multiplied by X before producing Q, K, and V. The multiplied result between query (Q) and key (K) yields an attention score matrix. This can also be seen as a database where we use the query and keys in order to find out how much various items are related in terms of numeric evaluation. Multiplication of the attention score and the V matrix produces the final result of this type of attention mechanism. The main reason for it being called **self-attention** is because of its unified input X; Q, K, and V are computed from X. You can see all this depicted in the following diagram:

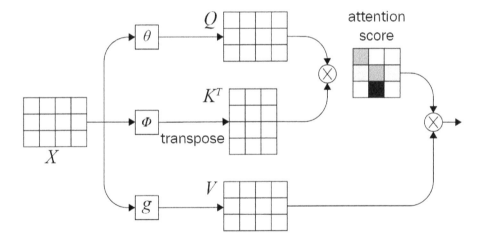

Figure 1.15 – Mathematical representation for the attention mechanism (Image inspired from https://blogs.oracle.com/datascience/multi-head-self-attention-in-nlp)

A scaled dot-product attention mechanism is very similar to a self-attention (dot-product) mechanism except it uses a scaling factor. The multi-head part, on the other hand, ensures the model is capable of looking at various aspects of input at all levels. Transformer models attend to encoder annotations and the hidden values from past layers. The architecture of the Transformer model does not have a recurrent step-by-step flow; instead, it uses positional encoding in order to have information about the position of each token in the input sequence. The concatenated values of the embeddings (randomly initialized) and the fixed values of positional encoding are the input fed into the layers in the first encoder part and are propagated through the architecture, as illustrated in the following diagram:

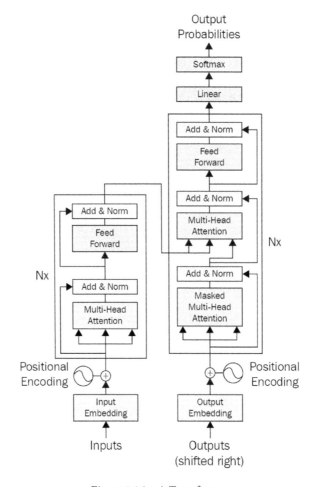

Figure 1.16 – A Transformer

The positional information is obtained by evaluating sine and cosine waves at different frequencies. An example of positional encoding is visualized in the following screenshot:

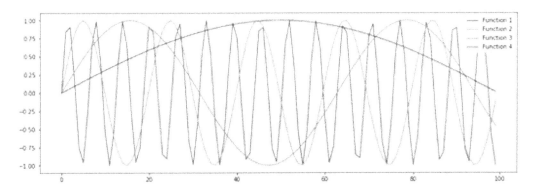

Figure 1.17 – Positional encoding (Image inspired from http://jalammar.github.io/illustrated-Transformer/)

A good example of performance on the Transformer architecture and the scaled dot-product attention mechanism is given in the following popular screenshot:

The	The
animal	animal
didn't	didn't
cross	cross
the	the
street	street
because	because
it	it
was	was
too	too
tired	tired
.	.

The	The
animal	animal
didn't	didn't
cross	cross
the	the
street	street
because	because
it	it
was	was
too	too
wide	wide
.	.

Figure 1.18 – Attention mapping for Transformers (Image inspired from https://ai.googleblog. com/2017/08/Transformer-novel-neural-network.html)

The word **it** refers to different entities in different contexts, as is seen from the preceding screenshot. Another improvement made by using a Transformer architecture is in parallelism. Conventional sequential recurrent models such as LSTMs and GRUs do not have such capabilities because they process the input token by token. Feed-forward layers, on the other hand, speed up a bit more because single matrix multiplication is far faster than a recurrent unit. Stacks of multi-head attention layers gain a better understanding of complex sentences. A good visual example of a multi-head attention mechanism is shown in the following screenshot:

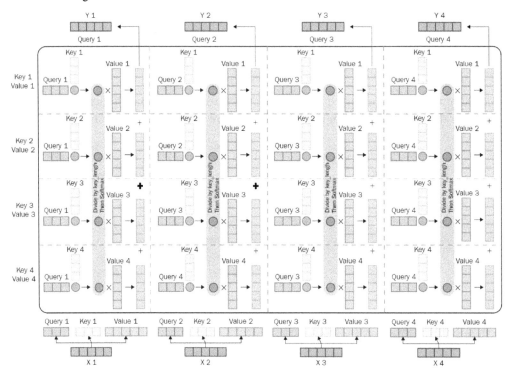

Figure 1.19 – Multi-head attention mechanism (Image inspired from https://imgur.com/gallery/FBQqrxw)

On the decoder side of the attention mechanism, a very similar approach to the encoder is utilized with small modifications. A multi-head attention mechanism is the same, but the output of the encoder stack is also used. This encoding is given to each decoder stack in the second multi-head attention layer. This little modification introduces the output of the encoder stack while decoding. This modification lets the model be aware of the encoder output while decoding and at the same time help it during training to have a better gradient flow over various layers. The final softmax layer at the end of the decoder layer is used to provide outputs for various use cases such as NMT, for which the original Transformer architecture was introduced.

This architecture has two inputs, noted as inputs and outputs (shifted right). One is always present (the inputs) in both training and inference, while the other is just present in training and in inference, which is produced by the model. The reason we do not use model predictions in inference is to stop the model from going too wrong by itself. But what does it mean? Imagine a neural translation model trying to translate a sentence from English to French—at each step, it makes a prediction for a word, and it uses that predicted word to predict the next one. But if it goes wrong at some step, all the following predictions will be wrong too. To stop the model from going wrong like this, we provide the correct words as a shifted-right version.

A visual example of a Transformer model is given in the following diagram. It shows a Transformer model with two encoders and two decoder layers. The **Add & Normalize** layer from this diagram adds and normalizes the input it takes from the **Feed Forward** layer:

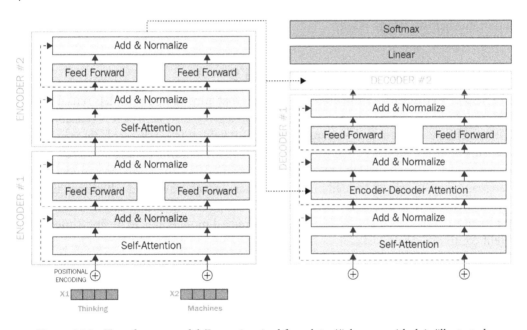

Figure 1.20 – Transformer model (Image inspired from http://jalammar.github.io/illustrated-Transformer/)

Another major improvement that is used by a Transformer-based architecture is based on a simple universal text-compression scheme to prevent unseen tokens on the input side. This approach, which takes place by using different methods such as byte-pair encoding and sentence-piece encoding, improves a Transformer's performance in dealing with unseen tokens. It also guides the model when the model encounters morphologically close tokens. Such tokens were unseen in the past and are rarely used in the training, and yet, an inference might be seen. In some cases, chunks of it are seen in training; the latter happens in the case of morphologically rich languages such as Turkish, German, Czech, and Latvian. For example, a model might see the word *training* but not *trainings*. In such cases, it can tokenize *trainings* as *training+s*. These two are commonly seen when we look at them as two parts.

Transformer-based models have quite common characteristics—for example, they are all based on this original architecture with differences in which steps they use and don't use. In some cases, minor differences are made—for example, improvements to the multi-head attention mechanism taking place.

Using TL with Transformers

TL is a field of **Artificial Intelligence** (**AI**) and ML that aims to make models reusable for different tasks—for example, a model trained on a given task such as *A* is reusable (fine-tuning) on a different task such as *B*. In an NLP field, this is achievable by using Transformer-like architectures that can capture the understanding of language itself by language modeling. Such models are called language models—they provide a model for the language they have been trained on. TL is not a new technique, and it has been used in various fields such as computer vision. ResNet, Inception, VGG, and EfficientNet are examples of such models that can be used as pre-trained models able to be fine-tuned on different computer-vision tasks.

Shallow TL using models such as *Word2vec*, *GloVe*, and *Doc2vec* is also possible in NLP. It is called *shallow* because there is no model behind this kind of TL and instead, the pre-trained vectors for words/tokens are utilized. You can use these token- or document-embedding models followed by a classifier or use them combined with other models such as RNNs instead of using random embeddings.

TL in NLP using Transformer models is also possible because these models can learn a language itself without any labeled data. Language modeling is a task used to train transferable weights for various problems. Masked language modeling is one of the methods used to learn a language itself. As with Word2vec's window-based model for predicting center tokens, in masked language modeling, a similar approach takes place, with key differences. Given a probability, each word is masked and replaced with a special token such as *[MASK]*. The language model (a Transformer-based model, in our case) must predict the masked words. Instead of using a window, unlike with Word2vec, a whole sentence is given, and the output of the model must be the same sentence with masked words filled.

One of the first models that used the Transformer architecture for language modeling is **BERT**, which is based on the encoder part of the Transformer architecture. Masked language modeling is accomplished by BERT by using the same method described before and after training a language model. BERT is a transferable language model for different NLP tasks such as token classification, sequence classification, or even question answering.

Each of these tasks is a fine-tuning task for BERT once a language model is trained. BERT is best known for its key characteristics on the base Transformer encoder model, and by altering these characteristics, different versions of it—small, tiny, base, large, and extra-large—are proposed. Contextual embedding enables a model to have the correct meaning of each word based on the context in which it is given—for example, the word *Cold* can have different meanings in two different sentences: *Cold-hearted killer* and *Cold weather*. The number of layers at the encoder part, the input dimension, the output embedding dimension, and the number of multi-head attention mechanisms are these key characteristics, as illustrated in the following screenshot:

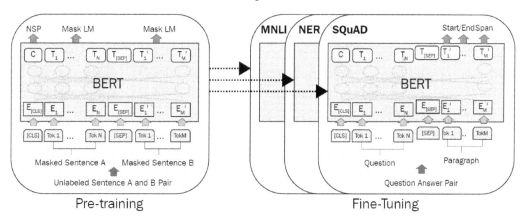

Figure 1.21 – Pre-training and fine-tuning procedures for BERT (Image inspired from J. Devlin et al., Bert: Pre-training of deep bidirectional Transformers for language understanding, 2018)

As you can see in *Figure 1.21*, the pre-training phase also consists of another objective known as **next-sentence prediction**. As we know, each document is composed of sentences followed by each other, and another important part of training for a model to grasp the language is to understand the relations of sentences to each other—in other words, whether they are related or not. To achieve these tasks, BERT introduced special tokens such as *[CLS]* and *[SEP]*. A *[CLS]* token is an initially meaningless token used as a start token for all tasks, and it contains all information about the sentence. In sequence-classification tasks such as NSP, a classifier on top of the output of this token (output position of *0*) is used. It is also useful in evaluating the sense of a sentence or capturing its semantics—for example, when using a Siamese BERT model, comparing these two *[CLS]* tokens for different sentences by a metric such as cosine-similarity is very helpful. On the other hand, *[SEP]* is used to distinguish between two sentences, and it is only used to separate two sentences. After pre-training, if someone aims to fine-tune BERT on a sequence-classification task such as sentiment analysis, which is a sequence-classification task, they will use a classifier on top of the output embedding of *[CLS]*. It is also notable that all TL models can be frozen during fine-tuning or freed; frozen means seeing all weights and biases inside the model as constants and stopping training on them. In the example of sentiment analysis, just the classifier will be trained, not the model if it is frozen.

Summary

With this, we now come to the end of the chapter. You should now have an understanding of the evolution of NLP methods and approaches, from BoW to Transformers. We looked at how to implement BoW-, RNN-, and CNN-based approaches and understood what Word2vec is and how it helps improve the conventional DL-based methods using shallow TL. We also looked into the foundation of the Transformer architecture, with BERT as an example. By the end of the chapter, we had learned about TL and how it is utilized by BERT.

At this point, we have learned basic information that is necessary to continue to the next chapters. We understood the main idea behind Transformer-based architectures and how TL can be applied using this architecture.

In the next section, we will see how it is possible to run a simple Transformer example from scratch. The related information about the installation steps will be given, and working with datasets and benchmarks is also investigated in detail.

References

- *Mikolov, T., Chen, K., Corrado, G. & Dean, J. (2013). Efficient estimation of word representations in vector space. arXiv preprint arXiv:1301.3781.*

- *Bahdanau, D., Cho, K. & Bengio, Y. (2014). Neural machine translation by jointly learning to align and translate. arXiv preprint arXiv:1409.0473.*

- *Pennington, J., Socher, R. & Manning, C. D. (2014, October). GloVe: Global vectors for word representation. In Proceedings of the 2014 conference on empirical methods in natural language processing (EMNLP) (pp. 1532-1543).*

- *Hochreiter, S. & Schmidhuber, J. (1997). Long short-term memory. Neural computation, 9(8), 1735-1780.*

- *Bengio, Y., Simard, P., & Frasconi, P. (1994). Learning long-term dependencies with gradient descent is difficult. IEEE transactions on neural networks, 5(2), 157-166.*

- *Cho, K., Van Merriënboer, B., Gulcehre, C., Bahdanau, D., Bougares, F., Schwenk, H. & Bengio, Y. (2014). Learning phrase representations using RNN encoder-decoder for statistical machine translation. arXiv preprint arXiv:1406.1078.*

- *Kim, Y. (2014). Convolutional neural networks for sentence classification. CoRR abs/1408.5882 (2014). arXiv preprint arXiv:1408.5882.*

- *Vaswani, A., Shazeer, N., Parmar, N., Uszkoreit, J., Jones, L., Gomez, A. N. & Polosukhin, I. (2017). Attention is all you need. arXiv preprint arXiv:1706.03762.*

- *Devlin, J., Chang, M. W., Lee, K. & Toutanova, K. (2018). Bert: Pre-training of deep bidirectional Transformers for language understanding. arXiv preprint arXiv:1810.04805.*

2
A Hands-On Introduction to the Subject

So far, we have had an overall look at the evolution of **Natural Language Processing** (**NLP**) using **Deep Learning** (**DL**)-based methods. We have learned some basic information about Transformer and their respective architecture. In this chapter, we are going to have a deeper look into how a transformer model can be used. Tokenizers and models, such as **Bidirectional Encoder Representations from Transformer** (**BERT**), will be described in more technical detail in this chapter with hands-on examples, including how to load a tokenizer/model and use community-provided pretrained models. But before using any specific model, we will understand the installation steps required to provide the necessary environment by using Anaconda. In the installation steps, installing libraries and programs on various operating systems such as Linux, Windows, and macOS will be covered. The installation of **PyTorch** and **TensorFlow**, in two versions of a **Central Processing Unit** (**CPU**) and a **Graphics Processing Unit** (**GPU**), is also shown. A quick jump into a **Google Colaboratory** (**Google Colab**) installation of the Transformer library is provided. There is also a section dedicated to using models in the PyTorch and TensorFlow frameworks.

The HuggingFace models repository is also another important part of this chapter, in which finding different models and steps to use various pipelines are discussed—for example, models such as **Bidirectional and Auto-Regressive Transformer (BART)**, BERT, and **TAble PArSing (TAPAS)** are detailed, with a glance at **Generative Pre-trained Transformer 2 (GPT-2)** text generation. However, this is purely an overview, and this part of the chapter relates to getting the environment ready and using pretrained models. No model training is discussed here as this is given greater significance in upcoming chapters.

After everything is ready and we have understood how to use the `Transformer` library for inference by community-provided models, the `datasets` library is described. Here, we look at loading various datasets, benchmarks, and using metrics. Loading a specific dataset and getting data back from it is one of the main areas we look at here. Cross-lingual datasets and how to use local files with the `datasets` library are also considered here. The `map` and `filter` functions are important functions of the `datasets` library in terms of model training and are also examined in this chapter.

This chapter is an essential part of the book because the `datasets` library is described in more detail here. It's also very important for you to understand how to use community-provided models and get the system ready for the rest of the book.

To sum all this up, we will cover the following topics in this chapter:

- Installing Transformer with Anaconda
- Working with language models and tokenizers
- Working with community-provided models
- Working with benchmarks and datasets
- Benchmarking for speed and memory

Technical requirements

You will need to install the libraries and software listed next. Although having the latest version is a plus, it is mandatory to install versions that are compatible with each other. For more information about the latest version installation for HuggingFace Transformer, take a look at their official web page at `https://huggingface.co/Transformer/installation.html`:

- Anaconda

- Transformer 4.0.0

- PyTorch 1.1.0

- TensorFlow 2.4.0

- Datasets 1.4.1

Finally, all the code shown in this chapter is available in this book's GitHub repository at `https://github.com/PacktPublishing/Mastering-Transformer/tree/main/CH02`.

Check out the following link to see the Code in Action video: `https://bit.ly/372ek48`

Installing Transformer with Anaconda

Anaconda is a distribution of the Python and R programming languages that makes package distribution and deployment easy for scientific computation. In this chapter, we will describe the installation of the `Transformer` library. However, it is also possible to install this library without the aid of Anaconda. The main motivation to use Anaconda is to explain the process more easily and moderate the packages used.

To start installing the related libraries, the installation of Anaconda is a mandatory step. Official guidelines provided by the Anaconda documentation offer simple steps to install it for common operating systems (macOS, Windows, and Linux).

Installation on Linux

Many distributions of Linux are available for users to enjoy, but among them, **Ubuntu** is one of the preferred ones. In this section, the steps to install Anaconda are covered for Linux. Proceed as follows:

1. Download the Anaconda installer for Linux from `https://www.anaconda.com/products/individual#Downloads` and go to the Linux section, as illustrated in the following screenshot:

Linux 🐧

Python 3.8

64-Bit (x86) Installer (529 MB)

64-Bit (Power8 and Power9) Installer (279 MB)

Figure 2.1 – Anaconda download link for Linux

2. Run a `bash` command to install it and complete the following steps:

3. Open the Terminal and run the following command:

```
bash Terminal./FilePath/For/Anaconda.sh
```

4. Press *Enter* to see the license agreement and press *Q* if you do not want to read it all, and then do the following:

5. Click **Yes** to agree.

6. Click **Yes** for the installer to always initialize the `conda` root environment.

7. After running a `python` command from the Terminal, you should see an Anaconda prompt after the Python version information.

8. You can access Anaconda Navigator by running an `anaconda-navigator` command from the Terminal. As a result, you will see the Anaconda **Graphical User Interface** (**GUI**) start loading the related modules, as shown in the following screenshot:

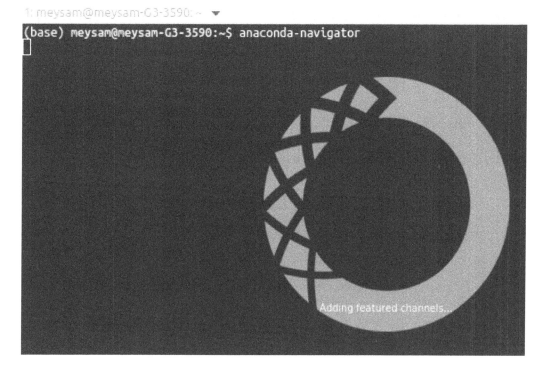

Figure 2.2 – Anaconda Navigator

Let's move on to the next section!

Installation on Windows

The following steps describe how you can install Anaconda on Windows operating systems:

1. Download the installer from `https://www.anaconda.com/products/` `individual#Downloads` and go to the Windows section, as illustrated in the following screenshot:

Figure 2.3 – Anaconda download link for Windows

2. Open the installer and follow the guide by clicking the **I Agree** button.

3. Select the location for installation, as illustrated in the following screenshot:

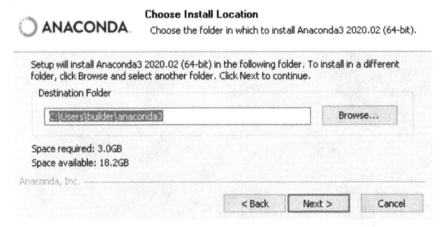

Figure 2.4 – Anaconda installer for Windows

4. Don't forget to check the **Add anaconda3 to my PATH environment variable** checkbox, as illustrated in the following screenshot. If you do not check this box, the Anaconda version of Python will not be added to the Windows environment variables, and you will not be able to directly run it with a python command from the Windows shell or the Windows command line:

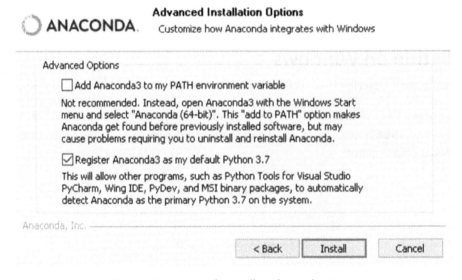

Figure 2.5 – Anaconda installer advanced options

5. Follow the rest of the installation instructions and finish the installation.

You should now be able to start Anaconda Navigator from the **Start** menu.

Installation on macOS

The following steps must be followed to install Anaconda on macOS:

1. Download the installer from `https://www.anaconda.com/products/individual#Downloads` and go to the macOS section, as illustrated in the following screenshot:

Figure 2.6 – Anaconda download link for macOS

2. Open the installer.

3. Follow the instructions and click the **Install** button to install macOS in a predefined location, as illustrated in the following screenshot. You can change the default directory, but this is not recommended:

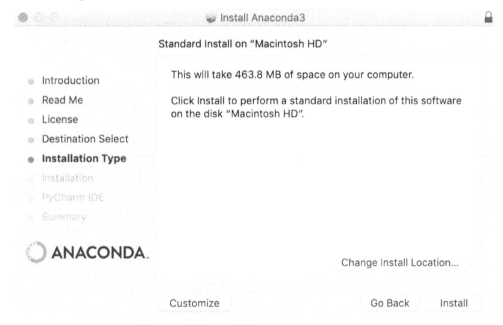

Figure 2.7 – Anaconda installer for macOS

Once you finish the installation, you should be able to access Anaconda Navigator.

Installing TensorFlow, PyTorch, and Transformer

The installation of TensorFlow and PyTorch as two major libraries that are used for DL can be made through `pip` or `conda` itself. `conda` provides a **Command-Line Interface (CLI)** for easier installation of these libraries.

For a clean installation and to avoid interrupting other environments, it is better to create a `conda` environment for the `huggingface` library. You can do this by running the following code:

```
conda create -n Transformer
```

This command will create an empty environment for installing other libraries. Once created, we will need to activate it, as follows:

```
conda activate Transformer
```

Installation of the `Transformer` library is easily done by running the following commands:

```
conda install -c conda-forge tensorflow
conda install -c conda-forge pytorch
conda install -c conda-forge Transformer
```

The `-c` argument in the `conda install` command lets Anaconda use additional channels to search for libraries.

Note that it is a requirement to have TensorFlow and PyTorch installed because the `Transformer` library uses both of these libraries. An additional note is the easy handling of CPU and GPU versions of TensorFlow by Conda. If you simply put –gpu after `tensorflow`, it will install the GPU version automatically. For installation of PyTorch through the `cuda` library (GPU version), you are required to have related libraries such as `cuda`, but `conda` handles this automatically and no further manual setup or installation is required. The following screenshot shows how `conda` automatically takes care of installing the PyTorch GPU version by installing the related `cudatoolkit` and `cudnn` libraries:

```
1: meysam@meysam-G3-3590: ~  ▼
(base) meysam@meysam-G3-3590:~$ conda install pytorch
Collecting package metadata (current_repodata.json): done
Solving environment: /
The environment is inconsistent, please check the package plan carefully
The following packages are causing the inconsistency:
  - defaults/linux-64::anaconda-navigator==1.9.12=py38_0
  - defaults/linux-64::ipykernel==5.3.4=py38h5ca1d4c_0
  - defaults/noarch::conda-verify==3.4.2=py_1
  - defaults/linux-64::notebook==6.1.4=py38_0
  - defaults/linux-64::nb_conda_kernels==2.2.3=py38_0
  - defaults/noarch::nbclient==0.5.1=py_0
  - conda-forge/linux-64::conda==4.9.2=py38h578d9bd_0
  - defaults/noarch::jupyter_client==6.1.7=py_0
  - defaults/linux-64::terminado==0.9.1=py38_0
  - defaults/linux-64::nbconvert==6.0.7=py38_0
  - defaults/linux-64::conda-build==3.18.11=py38_0
  - conda-forge/linux-64::widgetsnbextension==3.5.1=py38h32f6830_1
  - conda-forge/noarch::ipywidgets==7.5.1=pyh9f0ad1d_1
  - defaults/linux-64::_ipyw_jlab_nb_ext_conf==0.1.0=py38_0
  - defaults/linux-64::conda-package-handling==1.7.2=py38h03888b9_0
done

## Package Plan ##

  environment location: /home/meysam/anaconda3

  added / updated specs:
    - pytorch

The following packages will be downloaded:

    package                    |            build
    ---------------------------|-----------------
    _openmp_mutex-4.5          |            1_gnu           22 KB
    _pytorch_select-0.1        |            cpu_0            3 KB
    anyio-2.2.0                |   py38h06a4308_1          125 KB
    babel-2.9.1                |     pyhd3eb1b0_0          5.5 MB
    ca-certificates-2021.7.5   |       h06a4308_1          113 KB
    certifi-2021.5.30          |   py38h06a4308_0          138 KB
```

Figure 2.8 – Conda installing PyTorch and related cuda libraries

Note that all of these installations can also be done without `conda`, but the reason behind using Anaconda is its ease of use. In terms of using environments or installing GPU versions of TensorFlow or PyTorch, Anaconda works like magic and is a good time saver.

Installing using Google Colab

Even if the utilization of Anaconda saves time and is useful, in most cases, not everyone has such a good and reasonable computation resource available. Google Colab is a good alternative in such cases. Installation of the `Transformer` library in Colab is carried out with the following command:

```
!pip install Transformer
```

An exclamation mark before the statement makes the code run in a Colab shell, which is equivalent to running the code in the Terminal instead of running it using a Python interpreter. This will automatically install the `Transformer` library.

Working with language models and tokenizers

In this section, we will look at using the `Transformer` library with language models, along with their related **tokenizers**. In order to use any specified language model, we first need to import it. We will start with the BERT model provided by Google and use its pretrained version, as follows:

```
>>> from Transformer import BERTTokenizer
>>> tokenizer = \
BERTTokenizer.from_pretrained('BERT-base-uncased')
```

The first line of the preceding code snippet imports the BERT tokenizer, and the second line downloads a pretrained tokenizer for the BERT base version. Note that the uncased version is trained with uncased letters, so it does not matter whether the letters appear in upper- or lowercase. To test and see the output, you must run the following line of code:

```
>>> text = "Using Transformer is easy!"
>>> tokenizer(text)
```

This will be the output:

```
{'input_ids': [101, 2478, 19081, 2003, 3733, 999, 102], 'token_
type_ids': [0, 0, 0, 0, 0, 0, 0], 'attention_mask': [1, 1, 1,
1, 1, 1, 1]}
```

input_ids shows the token ID for each token, and token_type_ids shows the type of each token that separates the first and second sequence, as shown in the following screenshot:

```
0 0 0 0 0 0 0 0 0 1 1 1 1 1 1 1 1
| first sequence    | second sequence |
```

Figure 2.9 – Sequence separation for BERT

`attention_mask` is a mask of 0s and 1s that is used to show the start and end of a sequence for the transformer model in order to prevent unnecessary computations. Each tokenizer has its own way of adding special tokens to the original sequence. In the case of the BERT tokenizer, it adds a `[CLS]` token to the beginning and an `[SEP]` token to the end of the sequence, which can be seen by 101 and 102. These numbers come from the token IDs of the pretrained tokenizer.

A tokenizer can be used for both PyTorch- and TensorFlow-based `Transformer` models. In order to have output for each one, `pt` and `tf` keywords must be used in `return_tensors`. For example, you can use a tokenizer by simply running the following command:

```
>>> encoded_input = tokenizer(text, return_tensors="pt")
```

`encoded_input` has the tokenized text to be used by the PyTorch model. In order to run the model—for example, the BERT base model—the following code can be used to download the model from the `huggingface` model repository:

```
>>> from Transformer import BERTModel
>>> model = BERTModel.from_pretrained("BERT-base-uncased")
```

The output of the tokenizer can be passed to the downloaded model with the following line of code:

```
>>> output = model(**encoded_input)
```

This will give you the output of the model in the form of embeddings and cross-attention outputs.

When loading and importing models, you can specify which version of a model you are trying to use. If you simply put `TF` before the name of a model, the `Transformer` library will load the TensorFlow version of it. The following code shows how to load and use the TensorFlow version of BERT base:

```
from Transformer import BERTTokenizer, TFBERTModel
tokenizer = \
BERTTokenizer.from_pretrained('BERT-base-uncased')
model = TFBERTModel.from_pretrained("BERT-base-uncased")
text = " Using Transformer is easy!"
encoded_input = tokenizer(text, return_tensors='tf')
output = model(**encoded_input)
```

For specific tasks such as filling masks using language models, there are pipelines designed by `huggingface` that are ready to use. For example, a task of filling a mask can be seen in the following code snippet:

```
>>> from Transformer import pipeline
>>> unmasker = \
pipeline('fill-mask', model='BERT-base-uncased')
>>> unmasker("The man worked as a [MASK].")
```

This code will produce the following output, which shows the scores and possible tokens to be placed in the [MASK] token:

```
[{'score': 0.09747539460659027,  'sequence': 'the man worked
as a carpenter.',  'token': 10533,  'token_str': 'carpenter'},
{'score': 0.052383217960596085,  'sequence': 'the man worked
as a waiter.',  'token': 15610,  'token_str': 'waiter'},
{'score': 0.049627091735601425,  'sequence': 'the man worked
as a barber.',  'token': 13362,  'token_str': 'barber'},
{'score': 0.03788605332374573,  'sequence': 'the man worked
as a mechanic.',  'token': 15893,  'token_str': 'mechanic'},
{'score': 0.03768084570765495,  'sequence': 'the man worked as
a salesman.',  'token': 18968,  'token_str': 'salesman'}]
```

To get a neat view with pandas, run the following code:

```
>>> pd.DataFrame(unmasker("The man worked as a [MASK]."))
```

The result can be seen in the following screenshot:

	score	sequence	token	token_str
0	0.097475	the man worked as a carpenter.	10533	carpenter
1	0.052383	the man worked as a waiter.	15610	waiter
2	0.049627	the man worked as a barber.	13362	barber
3	0.037886	the man worked as a mechanic.	15893	mechanic
4	0.037681	the man worked as a salesman.	18968	salesman

Figure 2.10 – Output of the BERT mask filling

So far, we have learned how to load and use a pretrained BERT model and have understood the basics of tokenizers, as well as the difference between PyTorch and TensorFlow versions of the models. In the next section, we will learn to work with community-provided models by loading different models, reading the related information provided by the model authors and using different pipelines such as text-generation or **Question Answering** (**QA**) pipelines.

Working with community-provided models

Hugging Face has tons of community models provided by collaborators from large **Artificial Intelligence** (**AI**) and **Information Technology** (**IT**) companies such as Google and Facebook. There are also many interesting models that individuals and universities provide. Accessing and using them is also very easy. To start, you should visit the Transformer models directory available at their website (`https://huggingface.co/models`), as shown in the following screenshot:

Figure 2.11 – Hugging Face models repository

Apart from these models, there are also many good and useful datasets available for NLP tasks. To start using some of these models, you can explore them by keyword searches, or just specify your major NLP task and pipeline.

For example, we are looking for a table QA model. After finding a model that we are interested in, a page such as the following one will be available from the Hugging Face website (`https://huggingface.co/google/tapas-base-finetuned-wtq`):

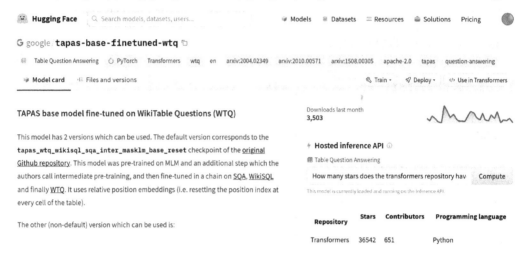

Figure 2.12 – TAPAS model page

On the right side, there is a panel where you can test this model. Note that this is a table QA model that can answer questions about a table provided to the model. If you ask a question, it will reply by highlighting the answer. The following screenshot shows how it gets input and provides an answer for a specific table:

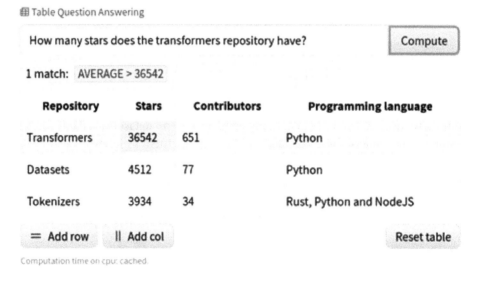

Figure 2.13 – Table QA using TAPAS

Each model has a page provided by the model authors that is also known as a **model card**. You can use the model by the examples provided in the model page. For example, you can visit the GPT-2 huggingface repository page and take a look at the example provided by the authors (https://huggingface.co/gpt2), as shown in the following screenshot:

How to use

You can use this model directly with a pipeline for text generation. Since the generation relies on some randomness, we set a seed for reproducibility:

```
>>> from transformers import pipeline, set_seed
>>> generator = pipeline('text-generation', model='gpt2')
>>> set_seed(42)
>>> generator("Hello, I'm a language model,", max_length=30, num_return_sequence

[{'generated_text': "Hello, I'm a language model, a language for thinking, a lan
 {'generated_text': "Hello, I'm a language model, a compiler, a compiler library
 {'generated_text': "Hello, I'm a language model, and also have more than a few
 {'generated_text': "Hello, I'm a language model, a system model. I want to know
 {'generated_text': 'Hello, I\'m a language model, not a language model"\n\nThe
```

Figure 2.14 – Text-generation code example from the Hugging Face GPT-2 page

Using pipelines is recommended because all the dirty work is taken care of by the Transformer library. As another example, let's assume you need an out-of-the-box zero-shot classifier. The following code snippet shows how easy it is to implement and use such a pretrained model:

```
>>> from Transformer import pipeline
>>> classifier = pipeline("zero-shot-classification",
model="facebook/bart-large-mnli")
>>> sequence_to_classify = "I am going to france."
>>> candidate_labels = ['travel', 'cooking', 'dancing']
>>> classifier(sequence_to_classify, candidate_labels)
```

The preceding code will provide the following result:

```
{'labels': ['travel', 'dancing', 'cooking'],
 'scores': [0.9866883158683777, 0.007197578903287649,
 0.006114077754318714], 'sequence': 'I am going to france.'}
```

We are done with the installation and the `hello-world` application part. So far, we have introduced the installation process, completed the environment settings, and experienced the first transformer pipeline. In the next part, we will introduce the `datasets` library, which will be our essential utility in the experimental chapters to come.

Working with benchmarks and datasets

Before introducing the `datasets` library, we'd better talk about important benchmarks such as **General Language Understanding Evalution (GLUE)**, **Cross-lingual TRansfer Evaluation of Multilingual Encoders (XTREME)**, and **Stanford Question Answering Dataset (SquAD)**. **Benchmarking** is especially critical for transferring learnings within multitask and multilingual environments. In NLP, we mostly focus on a particular metric that is a performance score on a certain task or dataset. Thanks to the `Transformer` library, we are able to transfer what we have learned from a particular task to a related task, which is called **Transfer Learning (TL)**. By transferring representations between related problems, we are able to train general-purpose models that share common linguistic knowledge across tasks, also known as **Multi-Task Learning (MTL)**. Another aspect of TL is to transfer knowledge across natural languages (multilingual models).

Important benchmarks

In this part, we will introduce the important benchmarks that are widely used by transformer-based architectures. These benchmarks exclusively contribute a lot to MTL and to multilingual and zero-shot learning, including many challenging tasks. We will look at the following benchmarks:

- **GLUE**
- **SuperGLUE**
- **XTREME**
- **XGLUE**
- **SQuAD**

For the sake of using fewer pages, we give details of the tasks for only the GLUE benchmark, so let's look at this benchmark first.

GLUE benchmark

Recent studies addressed the fact that multitask training approaches can achieve better results than single-task learning as a particular model for a task. In this direction, the **GLUE** benchmark has been introduced for MTL, which is a collection of tools and datasets for evaluating the performance of MTL models across a list of tasks. It offers a public leaderboard for monitoring submission performance on the benchmark, along with a single-number metric summarizing 11 tasks. This benchmark includes many sentence-understanding tasks that are based on existing tasks covering various datasets of differing size, text type, and difficulty levels. The tasks are categorized under three types, outlined as follows:

- **Single-sentence tasks**

- **CoLA**: The **Corpus of Linguistic Acceptability** dataset. This task consists of English acceptability judgments drawn from articles on linguistic theory.

- **SST-2**: The **Stanford Sentiment Treebank** dataset. This task includes sentences from movie reviews and human annotations of their sentiment with pos/neg labels.

- **Similarity and paraphrase tasks**

- **MRPC**: The **Microsoft Research Paraphrase Corpus** dataset. This task looks at whether the sentences in a pair are semantically equivalent.

- **QQP**: The **Quora Question Pairs** dataset. This task decides whether a pair of questions is semantically equivalent.

- **STS-B**: The **Semantic Textual Similarity Benchmark** dataset. This task is a collection of sentence pairs drawn from news headlines, with a similarity score between 1 and 5.

- **Inference tasks**

- **MNLI**: The **Multi-Genre Natural Language Inference** corpus. This is a collection of sentence pairs with textual entailment. The task is to predict whether the text entails a hypothesis (entailment), contradicts the hypothesis (contradiction), or neither (neutral).

- **QNLI**: **Question Natural Language Inference** dataset. This is a converted version of SquAD. The task is to check whether a sentence contains the answer to a question.

- **RTE**: The **Recognizing Textual Entailment** dataset. This is a task of textual entailment challenges to combine data from various sources. This dataset is similar to the previous QNLI dataset, where the task is to check whether a first text entails a second one.

- **WNLI**: The **Winograd Natural Language Inference** schema challenge. This is originally a pronoun resolution task linking a pronoun and a phrase in a sentence. GLUE converted the problem into sentence-pair classification, as detailed next.

SuperGLUE benchmark

Like Glue, **SuperGLUE** is a new benchmark styled with a new set of more difficult language-understanding tasks and offers a public leaderboard of around currently eight language tasks, drawing on existing data, associated with a single-number performance metric like that of GLUE. The motivation behind it is that as of writing this book, the current state-of-the-art GLUE Score (90.8) surpasses human performance (87.1). Thus, SuperGLUE provides a more challenging and diverse task toward general-purpose, language-understanding technologies.

You can access both GLUE and SuperGLUE benchmarks at gluebenchmark.com.

XTREME benchmark

In recent years, NLP researchers have increasingly focused on learning general-purpose representations rather than a single task that can be applied to many related tasks. Another way of building a general-purpose language model is by using multilingual tasks. It has been observed that recent multilingual models such as **Multilingual BERT** (**mBERT**) and XLM-R pretrained on massive amounts of multilingual corpora have performed better when transferring them to other languages. Thus, the main advantage here is that cross-lingual generalization enables us to build successful NLP applications in resource-poor languages through zero-shot cross-lingual transfer.

In this direction, the **XTREME** benchmark has been designed. It currently includes around 40 different languages belonging to 12 language families and includes 9 different tasks that require reasoning for various levels of syntax or semantics. However, it is still challenging to scale up a model to cover over 7,000 world languages and there exists a trade-off between language coverage and model capability. Please check out the following link for more details on this: https://sites.research.google/xtreme.

XGLUE benchmark

XGLUE is another cross-lingual benchmark to evaluate and improve the performance of cross-lingual pretrained models for **Natural Language Understanding** (NLU) and **Natural Language Generation** (NLG). It originally consisted of 11 tasks over 19 languages. The main difference from XTREME is that the training data is only available in English for each task. This forces the language models to learn only from the textual data in English and transfer this knowledge to other languages, which is called zero-shot cross-lingual transfer capability. The second difference is that it has tasks for NLU and NLG at the same time. Please check out the following link for more details on this: `https://microsoft.github.io/XGLUE/`.

SQuAD benchmark

SQuAD is a widely used QA dataset in the NLP field. It provides a set of QA pairs to benchmark the reading comprehension capabilities of NLP models. It consists of a list of questions, a reading passage, and an answer annotated by crowdworkers on a set of Wikipedia articles. The answer to the question is a span of text from the reading passage. The initial version, SQuAD1.1, doesn't have an unanswerable option where the datasets are collected, so each question has an answer to be found somewhere in the reading passage. The NLP model is forced to answer the question even if this appears impossible. SQuAD2.0 is an improved version, whereby the NLP models must not only answer questions when possible, but should also abstain from answering when it is impossible to answer. SQuAD2.0 contains 50,000 unanswerable questions written adversarially by crowdworkers to look similar to answerable ones. Additionally, it also has 100,000 questions taken from SQuAD1.1.

Accessing the datasets with an Application Programming Interface

The `datasets` library provides a very efficient utility to load, process, and share datasets with the community through the Hugging Face hub. As with TensorFlow datasets, it makes it easier to download, cache, and dynamically load the sets directly from the original dataset host upon request. The library also provides evaluation metrics along with the data. Indeed, the hub does not hold or distribute the datasets. Instead, it keeps all information about the dataset, including the owner, preprocessing script, description, and download link. We need to check whether we have permission to use the datasets under their corresponding license. To see other features, please check the `dataset_infos.json` and `DataSet-Name.py` files of the corresponding dataset under the GitHub repository, at `https://github.com/huggingface/datasets/tree/master/datasets`.

Let's start by installing the `dataset` library, as follows:

```
pip install datasets
```

The following code automatically loads the `cola` dataset using the Hugging Face hub. The `datasets.load_dataset()` function downloads the loading script from the actual path if the data is not cached already:

```
from datasets import load_dataset
cola = load_dataset('glue', 'cola')
cola['train'][25:28]
```

> **Important note**
>
> Reusability of the datasets: As you rerun the code a couple of times, the `datasets` library starts caching your loading and manipulation request. It first stores the dataset and starts caching your operations on the dataset, such as splitting, selection, and sorting. You will see a warning message such as **reusing dataset xtreme (/home/savas/.cache/huggingface/dataset...)** or **loading cached sorted....**

In the preceding example, we downloaded the `cola` dataset from the GLUE benchmark and selected a few examples from the `train` split of it.

Currently, there are 661 NLP datasets and 21 metrics for diverse tasks, as the following code snippet shows:

```
from pprint import pprint
from datasets import list_datasets, list_metrics
all_d = list_datasets()
metrics = list_metrics()
print(f"{len(all_d)} datasets and {len(metrics)} metrics exist
in the hub\n")
pprint(all_d[:20], compact=True)
pprint(metrics, compact=True)
```

This is the output:

```
661 datasets and 21 metrics exist in the hub.
['acronym_identification', 'ade_corpus_v2', 'adversarial_qa',
 'aeslc', 'afrikaans_ner_corpus', 'ag_news', 'ai2_arc', 'air_
 dialogue', 'ajgt_twitter_ar', 'allegro_reviews', 'allocine',
```

```
'alt', 'amazon_polarity', 'amazon_reviews_multi', 'amazon_us_
reviews', 'ambig_qa', 'amttl', 'anli', 'app_reviews', 'aqua_
rat']
['accuracy', 'BERTscore', 'bleu', 'bleurt', 'comet', 'coval',
'f1', 'gleu', 'glue', 'indic_glue', 'meteor', 'precision',
'recall', 'rouge', 'sacrebleu', 'sari', 'seqeval', 'squad',
'squad_v2', 'wer', 'xnli']
```

A dataset might have several configurations. For instance, GLUE, as an aggregated benchmark, has many subsets, such as CoLA, SST-2, and MRPC, as we mentioned before. To access each GLUE benchmark dataset, we pass two arguments, where the first is `glue` and the second is a particular dataset of its example dataset (`cola` or `sst2`) that can be chosen. Likewise, the Wikipedia dataset has several configurations provided for several languages.

A dataset comes with the `DatasetDict` object, including several `Dataset` instances. When the split selection (`split='...'`) is used, we get `Dataset` instances. For example, the CoLA dataset comes with `DatasetDict`, where we have three splits: *train*, *validation*, and *test*. While train and validation datasets include two labels (`1` for acceptable, `0` for unacceptable), the label value of test split is `-1`, which means no-label.

Let's see the structure of the CoLA dataset object, as follows:

```
>>> cola = load_dataset('glue', 'cola')
>>> cola
DatasetDict({
train: Dataset({
features: ['sentence', 'label', 'idx'],
        num_rows: 8551 })
validation: Dataset({
features: ['sentence', 'label', 'idx'],
        num_rows: 1043 })
test: Dataset({
        features: ['sentence', 'label', 'idx'],
        num_rows: 1063  })
})
cola['train'][12]
{'idx': 12, 'label':1,'sentence':'Bill rolled out of the
room.'}
>>> cola['validation'][68]
```

```
{'idx': 68, 'label': 0, 'sentence': 'Which report that John was
incompetent did he submit?'}
>>> cola['test'][20]
{'idx': 20, 'label': -1, 'sentence': 'Has John seen Mary?'}
```

The dataset object has some additional metadata information that might be helpful for us: split, description, citation, homepage, license, and info. Let's run the following code:

```
>>> print("1#",cola["train"].description)
>>> print("2#",cola["train"].citation)
>>> print("3#",cola["train"].homepage)
1# GLUE, the General Language Understanding Evaluation
benchmark(https://gluebenchmark.com/) is a collection of
resources for training,evaluating, and analyzing natural
language understanding systems.2# @article{warstadt2018neural,
title={Neural Network Acceptability Judgments},
author={Warstadt, Alex and Singh, Amanpreet and Bowman,
Samuel R},  journal={arXiv preprint arXiv:1805.12471},
year={2018}}@inproceedings{wang2019glue,  title={{GLUE}:
A Multi-Task Benchmark and Analysis Platform for Natural
Language Understanding},  author={Wang, Alex and Singh,
Amanpreet and Michael, Julian and Hill, Felix and Levy, Omer
and Bowman, Samuel R.},  note={In the Proceedings of ICLR.},
year={2019}}3# https://nyu-mll.github.io/CoLA/
```

The GLUE benchmark provides many datasets, as mentioned previously. Let's download the MRPC dataset, as follows:

```
>>> mrpc = load_dataset('glue', 'mrpc')
```

Likewise, to access other GLUE tasks, we will change the second parameter, as follows:

```
>>> load_dataset('glue', 'XYZ')
```

In order to apply a sanity check of data availability, run the following piece of code:

```
>>> glue=['cola', 'sst2', 'mrpc', 'qqp', 'stsb', 'mnli',
        'mnli_mismatched', 'mnli_matched', 'qnli', 'rte',
        'wnli', 'ax']
>>> for g in glue:
        _=load_dataset('glue', g)
```

XTREME (working with a cross-lingual dataset) is another popular cross-lingual dataset that we already discussed. Let's pick the MLQA example from the XTREME set. MLQA is a subset of the XTREME benchmark, which is designed for assessing the performance of cross-lingual QA models. It includes about 5,000 extractive QA instances in the SQuAD format across seven languages, which are English, German, Arabic, Hindi, Vietnamese, Spanish, and Simplified Chinese.

For example, MLQA.en.de is an English-German QA example dataset and can be loaded as follows:

```
>>> en_de = load_dataset('xtreme', 'MLQA.en.de')
>>> en_de \
DatasetDict({
test: Dataset({features: ['id', 'title', 'context', 'question',
'answers'], num_rows: 4517
}) validation: Dataset({ features: ['id', 'title', 'context',
'question', 'answers'], num_rows: 512})})
```

It could be more convenient to view it within a pandas DataFrame, as follows:

```
>>> import pandas as pd
>>> pd.DataFrame(en_de['test'][0:4])
```

Here is the output of the preceding code:

	answers	context	id	question	title
0	{'answer_start': [31], 'text': ['cell']}	An established or immortalized cell line has a...	037e8929e7e4d2f949ffbabd10f0f860499ff7c9	Woraus besteht die Linie?	Cell culture
1	{'answer_start': [232], 'text': ['1885']}	The 19th-century English physiologist Sydney R...	4b36724f3cbde7c287bde512ff09194cbba7f932	Wann hat Roux etwas von seiner Medullarplatte ...	Cell culture
2	{'answer_start': [131], 'text': ['TRIPS']}	After the Uruguay round, the GATT became the b...	13e58403df16d88b0e2c665953e89575704942d4	Was muss ratifiziert werden, wenn ein Land ger...	TRIPS Agreement

Figure 2.15 – English-German cross-lingual QA dataset

Data manipulation with the datasets library

Datasets come with many dictionaries of subsets, where the split parameter is used to decide which subset(s) or portion of the subset is to be loaded. If this is none by default, it will return a dataset dictionary of all subsets (train, test, validation, or any other combination). If the split parameter is specified, it will return a single dataset rather than a dictionary. For the following example, we retrieve a train split of the cola dataset only:

```
>>> cola_train = load_dataset('glue', 'cola', split ='train')
```

We can get a mixture of train and validation subsets, as follows:

```
>>> cola_sel = load_dataset('glue', 'cola', split =
'train[:300]+validation[-30:]')
```

The split expression means that the first 300 examples of train and the last 30 examples of validation are obtained as cola_sel.

We can apply different combinations, as shown in the following split examples:

- The first 100 examples from train and validation, as shown here:

  ```
  split='train[:100]+validation[:100]'
  ```

- 50% of train and the last 30% of validation, as shown here:

  ```
  split='train[:50%]+validation[-30%:]'
  ```

- The first 20% of train and the examples in the slice [30:50] from validation, as shown here:

  ```
  split='train[:20%]+validation[30:50]'
  ```

Sorting, indexing, and shuffling

The following execution calls the sort() function of the cola_sel object. We see the first 15 and the last 15 labels:

```
>>> cola_sel.sort('label')['label'][:15]
[0, 0, 0, 0, 0, 0, 0, 0, 0, 0, 0, 0, 0, 0, 0]
>>> cola_sel.sort('label')['label'][-15:]
[1, 1, 1, 1, 1, 1, 1, 1, 1, 1, 1, 1, 1, 1, 1]
```

We are already familiar with Python slicing notation. Likewise, we can also access several rows using similar slice notation or with a list of indices, as follows:

```
>>> cola_sel[6,19,44]
{'idx': [6, 19, 44],
 'label': [1, 1, 1],
  'sentence':['Fred watered the plants flat.',
   'The professor talked us into a stupor.',
   'The trolley rumbled through the tunnel.']}
```

We shuffle the dataset as follows:

```
>>> cola_sel.shuffle(seed=42)[2:5]
{'idx': [159, 1022, 46],
 'label': [1, 0, 1],
 'sentence': ['Mary gets depressed if she listens to the
Grateful Dead.',
 'It was believed to be illegal by them to do that.',
 'The bullets whistled past the house.']}
```

> **Important note**
> Seed value: When shuffling, we need to pass a seed value to control the randomness and achieve a consistent output between the author and the reader.

Caching and reusability

Using cached files allows us to load large datasets by means of memory mapping (if datasets fit on the drive) by using a fast backend. Such smart caching helps in saving and reusing the results of operations executed on the drive. To see cache logs with regard to the dataset, run the following code:

```
>>> cola_sel.cache_files
[{'filename': '/home/savas/.cache/huggingface...,'skip':
0,  'take': 300}, {'filename': '/home/savas/.cache/
huggingface...','skip': 1013,  'take': 30}]
```

Dataset filter and map function

We might want to work with a specific selection of a dataset. For instance, we can retrieve sentences only, including the term kick in the cola dataset, as shown in the following execution. The datasets.Dataset.filter() function returns sentences including kick where an anonymous function and a lambda keyword are applied:

```
>>> cola_sel = load_dataset('glue', 'cola',
split='train[:100%]+validation[-30%:]')
>>> cola_sel.filter(lambda s: "kick" in s['sentence'])
["sentence"][:3]
['Jill kicked the ball from home plate to third base.', 'Fred
kicked the ball under the porch.', 'Fred kicked the ball behind
the tree.']
```

The following filtering is used to get positive (acceptable) examples from the set:

```
>>> cola_sel.filter(lambda s: s['label']== 1 )["sentence"][:3]
["Our friends won't buy this analysis, let alone the next one
we propose.",
"One more pseudo generalization and I'm giving up.",
"One more pseudo generalization or I'm giving up."]
```

In some cases, we might not know the integer code of a class label. Suppose we have many classes, and the code of the culture class is hard to remember out of 10 classes. Instead of giving integer code 1 in our preceding example, which is the code for acceptable, we can pass an acceptable label to the str2int() function, as follows:

```
>>> cola_sel.filter(lambda s: s['label']== cola_sel.
features['label'].str2int('acceptable'))["sentence"][:3]
```

This produces the same output as with the previous execution.

Processing data with the map function

The datasets.Dataset.map() function iterates over the dataset, applying a processing function to each example in the set, and modifies the content of the examples. The following execution shows a new 'len' feature being added that denotes the length of a sentence:

```
>>> cola_new=cola_sel.map(lambda e:{'len': len(e['sentence'])})
>>> pd.DataFrame(cola_new[0:3])
```

This is the output of the preceding code snippet:

	idx	label	len	sentence
0	0	1	71	Our friends won't buy this analysis, let alone...
1	1	1	49	One more pseudo generalization and I'm giving up.
2	2	1	48	One more pseudo generalization or I'm giving up.

Figure 2.16 – Cola dataset an with additional column

As another example, the following piece of code cut the sentence after 20 characters. We do not create a new feature, but instead update the content of the sentence feature, as follows:

```
>>> cola_cut=cola_new.map(lambda e: {'sentence': e['sentence']
[:20]+ '_'})
```

The output is shown here:

	idx	label	len	sentence
0	0	1	71	Our friends won't bu_
1	1	1	49	One more pseudo gene_
2	2	1	48	One more pseudo gene_

Figure 2.17 – Cola dataset with an update

Working with local files

To load a dataset from local files in a **Comma-Separated Values (CSV)**, **Text (TXT)**, or **JavaScript Object Notation (JSON)** format, we pass the file type (csv, text, or json) to the generic load_dataset() loading script, as shown in the following code snippet. Under the ../data/ folder, there are three CSV files (a.csv, b.csv, and c.csv), which are randomly selected toy examples from the SST-2 dataset. We can load a single file, as shown in the data1 object, merge many files, as in the data2 object, or make dataset splits, as in data3:

```
from datasets import load_dataset
data1 = load_dataset('csv', data_files='../data/a.csv',
delimiter="\t")
data2 = load_dataset('csv', data_files=['../data/a.csv','../
data/b.csv', '../data/c.csv'], delimiter="\t")
data3 = load_dataset('csv', data_files={'train':['../
data/a.csv','../data/b.csv'], 'test':['../data/c.csv']},
delimiter="\t")
```

In order to get the files in other formats, we pass json or text instead of csv, as follows:

```
>>> data_json = load_dataset('json', data_files='a.json')
>>> data_text = load_dataset('text', data_files='a.txt')
```

So far, we have discussed how to load, handle, and manipulate datasets that are either already hosted in the hub or are on our local drive. Now, we will study how to prepare datasets for transformer model training.

Preparing a dataset for model training

Let's start with the tokenization process. Each model has its own tokenization model that is trained before the actual language model. We will discuss this in detail in the next chapter. To use a tokenizer, we should have installed the Transformer library. The following example loads the tokenizer model from the pretrained distilBERT-base-uncased model. We use map and an anonymous function with lambda to apply a tokenizer to each split in data3. If batched is selected True in the map function, it provides a batch of examples to the tokenizer function. The batch_size value is 1000 by default, which is the number of examples per batch passed to the function. If not selected, the whole dataset is passed as a single batch. The code can be seen here:

```
from Transformer import DistilBERTTokenizer
tokenizer = \ DistilBERTTokenizer.from_pretrained('distilBERT-base-uncased')
encoded_data3 = data3.map(lambda e: tokenizer( e['sentence'],
padding=True, truncation=True, max_length=12), batched=True,
batch_size=1000)
```

As shown in the following output, we see the difference between data3 and encoded_data3, where two additional features—attention_mask and input_ids—are added to the datasets accordingly. We already introduced these two features in the previous part in this chapter. Put simply, input_ids are the indices corresponding to each token in the sentence. They are expected features needed by the Trainer class of Transformer, which we will discuss in the next fine-tuning chapters.

We mostly pass several sentences at once (called a **batch**) to the tokenizer and further pass the tokenized batch to the model. To do so, we pad each sentence to the maximum sentence length in the batch or a particular maximum length specified by the max_length parameter—12 in this toy example. We also truncate longer sentences to fit that maximum length. The code can be seen in the following snippet:

```
>>> data3
DatasetDict({
```

```
train: Dataset({
    features: ['sentence','label'], num_rows: 199 })
test: Dataset({
    features: ['sentence','label'], num_rows: 100 })})
>>> encoded_data3
DatasetDict({
train: Dataset({
  features: ['attention_mask', 'input_ids', 'label',
'sentence'],
    num_rows: 199 })
test: Dataset({
features: ['attention_mask', 'input_ids', 'label', 'sentence'],
 num_rows: 100 })})
>>> pprint(encoded_data3['test'][12])
{'attention_mask': [1, 1, 1, 1, 1, 1, 1, 0, 0, 0, 0, 0],
'input_ids': [101, 2019, 5186, 16010, 2143, 1012, 102, 0, 0, 0,
0, 0], 'label': 0, 'sentence': 'an extremely unpleasant film .
'}
```

We are done with the datasets library. Up to this point, we have evaluated all aspects of datasets. We have covered GLUE-like benchmarking, where classification metrics are taken into consideration. In the next section, we will focus on how to benchmark computational performance for speed and memory rather than classification.

Benchmarking for speed and memory

Just comparing the classification performance of large models on a specific task or a benchmark turns out to be no longer sufficient. We must now take care of the computational cost of a particular model for a given environment (**Random-Access Memory (RAM)**, CPU, GPU) in terms of memory usage and speed. The computational cost of training and deploying to production for inference are two main values to be measured. Two classes of the Transformer library, PyTorchBenchmark and TensorFlowBenchmark, make it possible to benchmark models for both TensorFlow and PyTorch.

Before we start our experiment, we need to check our GPU capabilities with the following execution:

```
>>> import torch
>>> print(f"The GPU total memory is {torch.cuda.get_device_
```

```
properties(0).total_memory /(1024**3)} GB")
The GPU total memory is 2.94921875 GB
```

The output is obtained from NVIDIA GeForce GTX 1050 (3 **Gigabytes** (**GB**)). We need more powerful resources for an advanced implementation. The `Transformer` library currently only supports single-device benchmarking. When we conduct benchmarking on a GPU, we are expected to indicate on which GPU device the Python code will run, which is done by setting the `CUDA_VISIBLE_DEVICES` environment variable. For example, `export CUDA_VISIBLE_DEVICES=0.0` indicates that the first `cuda` device will be used.

In the code example that follows, two grids are explored. We compare four randomly selected pretrained BERT models, as listed in the `models` array. The second parameter to be observed is `sequence_lengths`. We keep the batch size as 4. If you have a better GPU capacity, you can extend the parameter search space with batch values in the range 4-64 and other parameters:

```
from Transformer import PyTorchBenchmark,
PyTorchBenchmarkArguments
models= ["BERT-base-uncased","distilBERT-base-
uncased","distilroBERTa-base", "distilBERT-base-german-cased"]
batch_sizes=[4]
sequence_lengths=[32,64, 128, 256,512]
args = PyTorchBenchmarkArguments(models=models, batch_
sizes=batch_sizes, sequence_lengths=sequence_lengths, multi_
process=False)
benchmark = PyTorchBenchmark(args)
```

> **Important note**
>
> Benchmarking for TensorFlow: The code examples are for PyTorch benchmarking in this part. For TensorFlow benchmarking, we simply use the `TensorFlowBenchmarkArguments` and `TensorFlowBenchmark` counterpart classes instead.

We are ready to conduct the benchmarking experiment by running the following code:

```
>>> results = benchmark.run()
```

This may take some time, depending on your CPU/GPU capacity and argument selection. If you face an out-of-memory problem for it, you should take the following actions to overcome this:

- Restart your kernel or your operating system.

- Delete all unnecessary objects in the memory before starting.

- Set a lower batch size, such as 2, or even 1.

The following output indicates the inference speed performance. Since our search space has four different models and five different sequence lengths, we see 20 rows in the results:

```
====================    INFERENCE - SPEED - RESULT    ====================
- - - - - - - - - - - - - - - - - - - - - - - - - - - - - - - - - - - - - - -
          Model Name           Batch Size      Seq Length      Time in s
- - - - - - - - - - - - - - - - - - - - - - - - - - - - - - - - - - - - - - -
      bert-base-uncased            4               32            0.021
      bert-base-uncased            4               64            0.031
      bert-base-uncased            4              128            0.057
      bert-base-uncased            4              256             0.12
      bert-base-uncased            4              512            0.269
   distilbert-base-uncased         4               32            0.007
   distilbert-base-uncased         4               64            0.011
   distilbert-base-uncased         4              128            0.021
   distilbert-base-uncased         4              256            0.044
   distilbert-base-uncased         4              512            0.095
     distilroberta-base            4               32            0.009
     distilroberta-base            4               64            0.014
     distilroberta-base            4              128            0.025
     distilroberta-base            4              256            0.053
     distilroberta-base            4              512            0.118
 distilbert-base-german-cased      4               32            0.007
 distilbert-base-german-cased      4               64            0.012
 distilbert-base-german-cased      4              128            0.021
 distilbert-base-german-cased      4              256            0.044
 distilbert-base-german-cased      4              512            0.095
- - - - - - - - - - - - - - - - - - - - - - - - - - - - - - - - - - - - - - -
```

Figure 2.18 – Inference speed performance

Likewise, we see the inference memory usage for 20 different scenarios, as follows:

```
====================    INFERENCE - MEMORY - RESULT    ====================
--------------------------------------------------------------------------
          Model Name        Batch Size     Seq Length    Memory in MB
--------------------------------------------------------------------------
      bert-base-uncased          4             32            1453
      bert-base-uncased          4             64            1487
      bert-base-uncased          4            128            1547
      bert-base-uncased          4            256            1661
      bert-base-uncased          4            512            1901
   distilbert-base-uncased       4             32            1908
   distilbert-base-uncased       4             64            1900
   distilbert-base-uncased       4            128            1900
   distilbert-base-uncased       4            256            1900
   distilbert-base-uncased       4            512            1900
      distilroberta-base         4             32            1907
      distilroberta-base         4             64            1900
      distilroberta-base         4            128            1900
      distilroberta-base         4            256            2098
      distilroberta-base         4            512            2492
distilbert-base-german-cased     4             32            2499
distilbert-base-german-cased     4             64            2492
distilbert-base-german-cased     4            128            2492
distilbert-base-german-cased     4            256            2491
distilbert-base-german-cased     4            512            2491
--------------------------------------------------------------------------
```

Figure 2.19 – Inference memory usage

To observe the memory usage across the parameters, we will plot them by using the `results` object that stores the statistics. The following execution will plot the time inference performance across models and sequence lengths:

```python
import matplotlib.pyplot as plt
plt.figure(figsize=(8,8))
t=sequence_lengths
models_perf=[list(results.time_inference_result[m]['result']
[batch_sizes[0]].values()) for m in models]
plt.xlabel('Seq Length')
plt.ylabel('Time in Second')
plt.title('Inference Speed Result')
plt.plot(t, models_perf[0], 'rs--', t, models_perf[1], 'g--.',
t, models_perf[2], 'b--^', t, models_perf[3], 'c--o')
plt.legend(models)
plt.show()
```

As shown in the following screenshot, two DistillBERT models showed close results and performed better than other two models. The BERT-based-uncased model performs poorly compared to the others, especially as the sequence length increases:

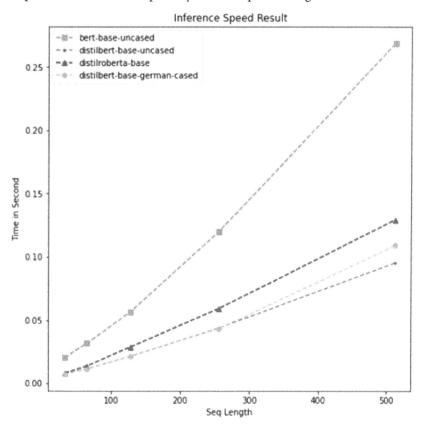

Figure 2.20 – Inference speed result

To plot the memory performance, you need to use the memory_inference_result result of the results object instead of time_inference_result, shown in the preceding code.

For more interesting benchmarking examples, please check out the following links:

- https://huggingface.co/Transformer/benchmarks.html
- https://github.com/huggingface/Transformer/blob/master/notebooks/05-benchmark.ipynb

Now that we are done with this section, we successfully completed this chapter. Congratulations on achieving the installation, running your first hello-world transformer program, working with the datasets library, and benchmarking!

Summary

In this chapter, we have covered a variety of introductory topics and also got our hands dirty with the `hello-world` transformer application. On the other hand, this chapter plays a crucial role in terms of applying what has been learned so far to the upcoming chapters. So, what has been learned so far? We took a first small step by setting the environment and system installation. In this context, the `anaconda` package manager helped us to install the necessary modules for the main operating systems. We also went through language models, community-provided models, and tokenization processes. Additionally, we introduced multitask (GLUE) and cross-lingual benchmarking (XTREME), which enables these language models to become stronger and more accurate. The `datasets` library was introduced, which facilitates efficient access to NLP datasets provided by the community. Finally, we learned how to evaluate the computational cost of a particular model in terms of memory usage and speed. Transformer frameworks make it possible to benchmark models for both TensorFlow and PyTorch.

The models that have been used in this section were already trained and shared with us by the community. Now, it is our turn to train a language model and disseminate it to the community.

In the next chapter, we will learn how to train a BERT language model as well as a tokenizer, and look at how to share them with the community.

Section 2: Transformer Models – From Autoencoding to Autoregressive Models

In this section, you will learn about the architecture of autoencoding models such as BERT and autoregressive models such as GPT. You will learn how to train, test, and fine-tune the models for a variety of natural language understanding and natural language generation problems. You will also learn how to share the models with the community and how to fine-tune other pre-trained language models shared by the community.

This section comprises the following chapters:

- *Chapter 3, Autoencoding Language Models*
- *Chapter 4, Autoregressive and Other Language Models*
- *Chapter 5, Fine-Tuning Language Models for Text Classification*
- *Chapter 6, Fine-Tuning Language Models for Token Classification*
- *Chapter 7, Text Representation*

3
Autoencoding Language Models

In the previous chapter, we looked at and studied how a typical Transformer model can be used by HuggingFace's Transformers. So far, all the topics have included how to use pre-defined or pre-built models and less information has been given about specific models and their training.

In this chapter, we will gain knowledge of how we can train autoencoding language models on any given language from scratch. This training will include pre-training and task-specific training of the models. First, we will start with basic knowledge about the BERT model and how it works. Then we will train the language model using a simple and small corpus. Afterward, we will look at how the model can be used inside any Keras model.

For an overview of what will be learned in this chapter, we will discuss the following topics:

- BERT – one of the autoencoding language models
- Autoencoding language model training for any language
- Sharing models with the community
- Understanding other autoencoding models
- Working with tokenization algorithms

Technical requirements

The technical requirements for this chapter are as follows:

- Anaconda
- Transformers >= 4.0.0
- PyTorch >= 1.0.2
- TensorFlow >= 2.4.0
- Datasets >= 1.4.1
- Tokenizers

Please also check the corresponding GitHub code of chapter 03:

https://github.com/PacktPublishing/Advanced-Natural-Language-Processing-with-Transformers/tree/main/CH03.

Check out the following link to see Code in Action Video: https://bit.ly/3i1ycdY

BERT – one of the autoencoding language models

Bidirectional Encoder Representations from Transformers, also known as **BERT**, was one of the first autoencoding language models to utilize the encoder Transformer stack with slight modifications for language modeling.

The BERT architecture is a multilayer Transformer encoder based on the Transformer original implementation. The Transformer model itself was originally for machine translation tasks, but the main improvement made by BERT is the utilization of this part of the architecture to provide better language modeling. This language model, after pretraining, is able to provide a global understanding of the language it is trained on.

BERT language model pretraining tasks

To have a clear understanding of the masked language modeling used by BERT, let's define it with more details. **Masked language modeling** is the task of training a model on input (a sentence with some masked tokens) and obtaining the output as the whole sentence with the masked tokens filled. But how and why does it help a model to obtain better results on downstream tasks such as classification? The answer is simple: if the model can do a cloze test (a linguistic test for evaluating language understanding by filling in blanks), then it has a general understanding of the language itself. For other tasks, it has been pretrained (by language modeling) and will perform better.

Here's an example of a cloze test:

George Washington was the first President of the ___ States.

It is expected that *United* should fill in the blank. For a masked language model, the same task is applied, and it is required to fill in the masked tokens. However, masked tokens are selected randomly from a sentence.

Another task that BERT is trained on is **Next Sentence Prediction** (**NSP**). This pretraining task ensures that BERT learns not only the relations of all tokens to each other in predicting masked ones but also helps it understand the relation between two sentences. A pair of sentences is selected and given to BERT with a *[SEP]* splitter token in between. It is also known from the dataset whether the second sentence comes after the first one or not.

The following is an example of NSP:

It is required from reader to fill the blank. Bitcoin price is way over too high compared to other altcoins.

In this example, the model is required to predict it as negative (the sentences are not related to each other).

These two pretraining tasks enable BERT to have an understanding of the language itself. BERT token embeddings provide a contextual embedding for each token. **Contextual embedding** means each token has an embedding that is completely related to the surrounding tokens. Unlike Word2Vec and such models, BERT provides better information for each token embedding. NSP tasks, on the other hand, enable BERT to have better embeddings for *[CLS]* tokens. This token, as was discussed in the first chapter, provides information about the whole input. *[CLS]* is used for classification tasks and in the pretraining part learns the overall embedding of the whole input. The following figure shows an overall look at the BERT model. *Figure 3.1* shows the respective input and output of the BERT model:

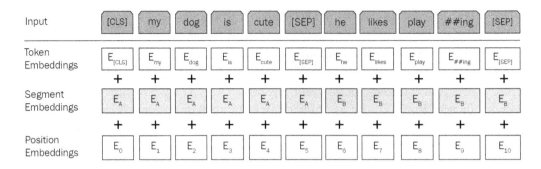

Figure 3.1 – The BERT model

Let's move on to the next section!

A deeper look into the BERT language model

Tokenizers are one of the most important parts of many NLP applications in their respective pipelines. For BERT, WordPiece tokenization is used. Generally, **WordPiece**, **SentencePiece**, and **BytePairEncoding** (**BPE**) are the three most widely known tokenizers, used by different Transformer-based architectures, which are also covered in the next sections. The main reason that BERT or any other Transformer-based architecture uses subword tokenization is the ability of such tokenizers to deal with unknown tokens.

BERT also uses positional encoding to ensure the position of the tokens is given to the model. If you recall from *Chapter 1*, *From Bag-of-Words to the Transformers*, BERT and similar models use non-sequential operations such as dense neural layers. Conventional models such as LSTM- and RNN-based models get the position by the order of the tokens in the sequence. In order to provide this extra information to BERT, positional encoding comes in handy.

Pretraining of BERT such as autoencoding models provides language-wise information for the model, but in practice, when dealing with different problems such as sequence classification, token classification, or question answering, different parts of the model output are used.

For example, in the case of a sequence classification task, such as sentiment analysis or sentence classification, it is proposed by the original BERT article that *[CLS]* embedding from the last layer must be used. However, there is other research that performs classification using different techniques using BERT (using average token embedding from all tokens, deploying an LSTM over the last layer, or even using a CNN on top of the last layer). The last *[CLS]* embedding for sequence classification can be used by any classifier, but the proposed, and the most common one, is a dense layer with an input size equal to the final token embedding size and an output size equal to the number of classes with a softmax activation function. Using sigmoid is also another alternative when the output could be multilabel and the problem itself is a multilabel classification problem.

To give you more detailed information about how BERT actually works, the following illustration shows an example of an NSP task. Note that the tokenization is simplified here for better understanding:

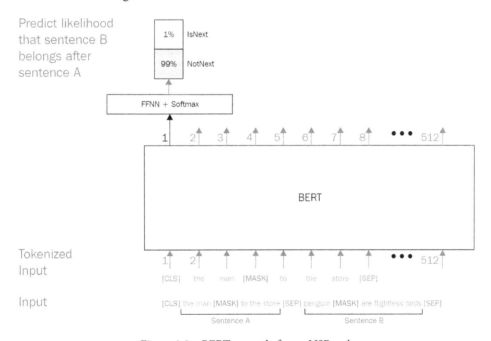

Figure 3.2 – BERT example for an NSP task

The BERT model has different variations, with different settings. For example, the size of the input is variable. In the preceding example, it is set to *512* and the maximum sequence size that model can get as input is *512*. However, this size includes special tokens, *[CLS]* and *[SEP]*, so it will be reduced to *510*. On the other hand, using WordPiece as a tokenizer yields subword tokens, and the sequence size before we can have fewer words, and after tokenization, the size will increase because the tokenizer breaks words into subwords if they are not commonly seen in the pretrained corpus.

The following figure shows an illustration of BERT for different tasks. For an NER task, the output of each token is used instead of *[CLS]*. In the case of question answering, the question and the answer are concatenated using the *[SEP]* delimiter token and the answer is annotated using *Start/End* and the *Span* output from the last layer. In this case, the *Paragraph* is the context that the *Question* is asked about it:

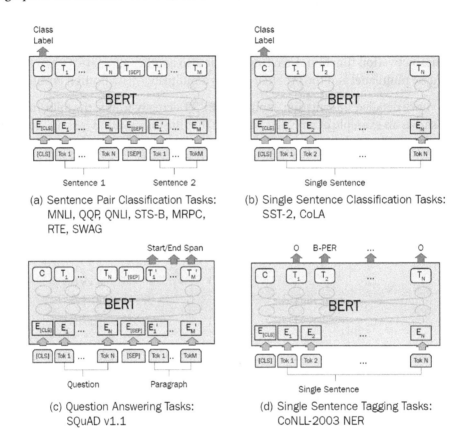

Figure 3.3 – BERT model for various NLP tasks

Regardless of all of these tasks, the most important ability of BERT is its contextual representation of text. The reason it is successful in various tasks is because of the Transformer encoder architecture that represents input in the form of dense vectors. These vectors can be easily transformed into output by very simple classifiers.

Up to this point, you have learned about BERT and how it works. You have learned detailed information on various tasks that BERT can be used for and the important points of this architecture.

In the next section, you will learn how you can pre-train BERT and use it after training.

Autoencoding language model training for any language

We have discussed how BERT works and that it is possible to use the pretrained version of it provided by the HuggingFace repository. In this section, you will learn how to use the HuggingFace library to train your own BERT.

Before we start, it is essential to have good training data, which will be used for the language modeling. This data is called the **corpus**, which is normally a huge pile of data (sometimes it is preprocessed and cleaned). This unlabeled corpus must be appropriate for the use case you wish to have your language model trained on; for example, if you are trying to have a special BERT for, let's say, the English language. Although there are tons of huge, good datasets, such as Common Crawl (https://commoncrawl.org/), we would prefer a small one for faster training.

The IMDB dataset of 50K movie reviews (available at https://www.kaggle.com/lakshmi25npathi/imdb-dataset-of-50k-movie-reviews) is a large dataset for sentiment analysis, but small if you use it as a corpus for training your language model:

1. You can easily download and save it in .txt format for language model and tokenizer training by using the following code:

    ```
    import pandas as pd
    imdb_df = pd.read_csv("IMDB Dataset.csv")
    reviews = imdb_df.review.to_string(index=None)
    with open("corpus.txt", "w") as f:
            f.writelines(reviews)
    ```

2. After preparing the corpus, the tokenizer must be trained. The `tokenizers` library provides fast and easy training for the WordPiece tokenizer. In order to train it on your corpus, it is required to run the following code:

    ```
    >>> from tokenizers import BertWordPieceTokenizer
    >>> bert_wordpiece_tokenizer =BertWordPieceTokenizer()
    >>> bert_wordpiece_tokenizer.train("corpus.txt")
    ```

3. This will train the tokenizer. You can access the trained vocabulary by using the `get_vocab()` function of the trained `tokenizer` object. You can get the vocabulary by using the following code:

    ```
    >>> bert_wordpiece_tokenizer.get_vocab()
    ```

The following is the output:

```
{'almod': 9111, 'events': 3710, 'bogart': 7647,
'slapstick': 9541, 'terrorist': 16811, 'patter': 9269,
'183': 16482, '##cul': 14292, 'sophie': 13109, 'thinki':
10265, 'tarnish': 16310, '##outh': 14729, 'peckinpah':
17156, 'gw': 6157, '##cat': 14290, '##eing': 14256,
'successfully': 12747, 'roomm': 7363, 'stalwart':
13347,...}
```

4. It is essential to save the tokenizer to be used afterwards. Using the `save_model()` function of the object and providing the directory will save the tokenizer vocabulary for further usage:

    ```
    >>> bert_wordpiece_tokenizer.save_model("tokenizer")
    ```

5. And you can reload it by using the `from_file()` function:

    ```
    >>> tokenizer = \ BertWordPieceTokenizer.from_
    file("tokenizer/vocab.txt")
    ```

6. You can use the tokenizer by following this example:

    ```
    >>> tokenized_sentence = \
    tokenizer.encode("Oh it works just fine")
    >>> tokenized_sentence.tokens
    ['[CLS]', 'oh', 'it', 'works', 'just', 'fine','[SEP]']
    ```

    The special tokens `[CLS]` and `[SEP]` will be automatically added to the list of tokens because BERT needs them for processing input.

7. Let's try another sentence using our tokenizer:

    ```
    >>> tokenized_sentence = \
    tokenizer.encode("ohoh i thougt it might be workingg
    well")
    ['[CLS]', 'oh', '##o', '##h', 'i', 'thoug', '##t', 'it',
    'might', 'be', 'working', '##g', 'well', '[SEP]']
    ```

8. Seems like a good tokenizer for noisy and misspelled text. Now that you have your tokenizer ready and saved, you can train your own BERT. The first step is to use `BertTokenizerFast` from the `Transformers` library. You are required to load the trained tokenizer from the previous step by using the following command:

```
>>> from Transformers import BertTokenizerFast
>>> tokenizer = \ BertTokenizerFast.from_
pretrained("tokenizer")
```

We have used `BertTokenizerFast` because it is suggested by the HuggingFace documentation. There is also `BertTokenizer`, which, according to the definition from the library documentation, is not implemented as fast as the fast version. In most of the pretrained models' documentations and cards, it is highly recommended to use the `BertTokenizerFast` version.

9. The next step is preparing the corpus for faster training by using the following command:

```
>>> from Transformers import LineByLineTextDataset
>>> dataset = \
LineByLineTextDataset(tokenizer=tokenizer,
                      file_path="corpus.txt",
                      block_size=128)
```

10. And it is required to provide a data collator for masked language modeling:

```
>>> from Transformers import
DataCollatorForLanguageModeling
>>> data_collator = DataCollatorForLanguageModeling(
                      tokenizer=tokenizer,
                      mlm=True,
                      mlm_probability=0.15)
```

The data collator gets the data and prepares it for the training. For example, the data collator above takes data and prepares it for masked language modeling with a probability of 0.15. The purpose of using such a mechanism is to do the preprocessing on the fly, which makes it possible to use fewer resources. On the other hand, it slows down the training process because each sample has to be preprocessed on the fly at training time.

11. The training arguments also provide information for the trainer in the training phase, which can be set by using the following command:

```
>>> from Transformers import TrainingArguments
>>> training_args = TrainingArguments(
                    output_dir="BERT",
                    overwrite_output_dir=True,
                    num_train_epochs=1,
                    per_device_train_batch_size=128)
```

12. We'll now make the BERT model itself, which we are going to use with the default configuration (the number of attention heads, Transformer encoder layers, and so on):

```
>>> from Transformers import BertConfig, BertForMaskedLM
>>> bert = BertForMaskedLM(BertConfig())
```

13. And the final step is to make a trainer object:

```
>>> from Transformers import Trainer
>>> trainer = Trainer(model=bert,
                    args=training_args,
                    data_collator=data_collator,
                    train_dataset=dataset)
```

14. Finally, you can train your language model using the following command:

```
>>> trainer.train()
```

It will show you a progress bar indicating the progress made in training:

█   [13/391 07:02 < 4:01:47, 0.03 it/s, Epoch 0.03/1]

Figure 3.4 – BERT model training progress

During the model training, a log directory called `runs` will be used to store the checkpoint in steps:

▼ 📁 runs
   › 📁 Mar18_20-51-26_cf17d0f459a7
   › 📁 Mar18_20-59-43_cf17d0f459a7

Figure 3.5 – BERT model checkpoints

15. After the training is finished, you can easily save the model using the following command:

```
>>> trainer.save_model("MyBERT")
```

Up to this point, you have learned how you can train BERT from scratch for any specific language that you desire. You've learned how to train the tokenizer and BERT model using the corpus you have prepared.

16. The default configuration that you provided for BERT is the most essential part of this training process, which defines the architecture of BERT and its hyperparameters. You can take a peek at these parameters by using the following code:

```
>>> from Transformers import BertConfig
>>> BertConfig()
```

The following is the output:

```
BertConfig {
    "attention_probs_dropout_prob": 0.1,
    "gradient_checkpointing": false,
    "hidden_act": "gelu",
    "hidden_dropout_prob": 0.1,
    "hidden_size": 768,
    "initializer_range": 0.02,
    "intermediate_size": 3072,
    "layer_norm_eps": 1e-12,
    "max_position_embeddings": 512,
    "model_type": "bert",
    "num_attention_heads": 12,
    "num_hidden_layers": 12,
    "pad_token_id": 0,
    "position_embedding_type": "absolute",
    "transformers_version": "4.4.2",
    "type_vocab_size": 2,
    "use_cache": true,
    "vocab_size": 30522
}
```

Figure 3.6 – BERT model configuration

If you wish to replicate **Tiny**, **Mini**, **Small**, **Base**, and relative models from the original BERT configurations (`https://github.com/google-research/bert`), you can change these settings:

	H=128	H=256	H=512	H=768
L=2	2/128 (BERT-Tiny)	2/256	2/512	2/768
L=4	4/128	4/256 (BERT-Mini)	4/512 (BERT-Small)	4/768
L=6	6/128	6/256	6/512	6/768
L=8	8/128	8/256	8/512 (BERT-Medium)	8/768
L=10	10/128	10/256	10/512	10/768
L=12	12/128	12/256	12/512	12/768 (BERT-Base)

Figure 3.7 – BERT model configurations (https://github.com/google-research/bert)

Note that changing these parameters, especially `max_position_embedding`, `num_attention_heads`, `num_hidden_layers`, `intermediate_size`, and `hidden_size`, directly affects the training time. Increasing them dramatically increases the training time for a large corpus.

17. For example, you can easily make a new configuration for a tiny version of BERT for faster training using the following code:

```
>>> tiny_bert_config = \ BertConfig(max_position_
embeddings=512, hidden_size=128,
            num_attention_heads=2,
            num_hidden_layers=2,
            intermediate_size=512)
>>> tiny_bert_config
```

The following is the result of the code:

```
BertConfig {
  "attention_probs_dropout_prob": 0.1,
  "gradient_checkpointing": false,
  "hidden_act": "gelu",
  "hidden_dropout_prob": 0.1,
  "hidden_size": 128,
  "initializer_range": 0.02,
  "intermediate_size": 512,
  "layer_norm_eps": 1e-12,
  "max_position_embeddings": 512,
  "model_type": "bert",
  "num_attention_heads": 2,
  "num_hidden_layers": 2,
  "pad_token_id": 0,
  "position_embedding_type": "absolute",
  "transformers_version": "4.4.2",
  "type_vocab_size": 2,
  "use_cache": true,
  "vocab_size": 30522
}
```

Figure 3.8 – Tiny BERT model configuration

18. By using the same method, we can make a tiny BERT model using this configuration:

```
>>> tiny_bert = BertForMaskedLM(tiny_bert_config)
```

19. And using the same parameters for training, you can train this tiny new BERT:

```
>>> trainer = Trainer(model=tiny_bert, args=training_
args,
                      data_collator=data_collator,
                      train_dataset=dataset)
>>> trainer.train()
```

The following is the output:

```
[ 9/391 00:17 < 15:43, 0.40 it/s, Epoch 0.02/1]
```

Figure 3.9 – Tiny BERT model configuration

It is clear that the training time is dramatically decreased, but you should be aware that this is a tiny version of BERT with fewer layers and parameters, which is not as good as BERT Base.

Up to this point, you have learned how to train your own model from scratch, but it is essential to note that using the `datasets` library is a better choice when dealing with datasets for training language models or leveraging it to perform task-specific training.

20. The BERT language model can also be used as an embedding layer combined with any deep learning model. For example, you can load any pretrained BERT model or your own version that has been trained in the previous step. The following code shows how you must load it to be used in a Keras model:

```
>>>  from Transformers import\
TFBertModel, BertTokenizerFast
>>>  bert = TFBertModel.from_pretrained(
"bert-base-uncased")
>>> tokenizer = BertTokenizerFast.from_pretrained(
"bert-base-uncased")
```

21. But you do not need the whole model; instead, you can access the layers by using the following code:

```
>>> bert.layers
[<Transformers.models.bert.modeling_tf_bert.
TFBertMainLayer at 0x7f72459b1110>]
```

22. As you can see, there is just a single layer from `TFBertMainLayer`, which you can access within your Keras model. But before using it, it is nice to test it and see what kind of output it provides:

```
>>> tokenized_text = tokenizer.batch_encode_plus(
                    ["hello how is it going with you",
                    "lets test it"],
                    return_tensors="tf",
                    max_length=256,
                    truncation=True,
                    pad_to_max_length=True)
>>> bert(tokenized_text)
```

The output is as follows:

```
TfBaseModelOutputWithPooling([['last_hidden_state',
                    <tf.Tensor: shape=(2, 256, 768), dtype=float32, numpy=
                    array([[[ 1.00471362e-01,  6.77026287e-02, -8.33595246e-02, ...,
                            -4.93304580e-01,  1.16539136e-01,  2.26647347e-01],
                           [ 3.23623657e-01,  3.70719165e-01,  6.14685774e-01, ...,
                            -6.27267540e-01,  3.79083097e-01,  7.05310702e-02],
                           [ 1.99532971e-01, -8.75509441e-01, -6.47868365e-02, ...,
                            -1.28077380e-02,  3.07651043e-01, -2.07325034e-02],
                           ...,
                           [-6.53299838e-02,  1.19046383e-01,  5.76846600e-01, ...,
                            -2.95460820e-01,  2.49744654e-02,  1.13964394e-01],
                           [-2.64715493e-01, -7.86386207e-02,  5.47280848e-01, ...,
                            -1.37515247e-01, -5.94691373e-02, -5.17928638e-02],
                           [-2.44958848e-01, -1.14799395e-01,  5.92173815e-01, ...,
                            -1.56882048e-01, -3.39757390e-02, -8.46135616e-02]],

                          [[ 2.94558890e-02,  2.30868042e-01,  2.92651534e-01, ...,
                            -1.30421281e-01,  1.89659461e-01,  4.68427837e-01],
                           [ 1.78523107e+00,  6.91360056e-01,  7.31509984e-01, ...,
                             2.89302200e-01,  5.36758840e-01, -1.54553086e-01],
                           [ 1.04596823e-01,  9.63676572e-02,  6.99661374e-02, ...,
                            -4.15922999e-01, -1.18989825e-01, -6.72240376e-01],
                           ...,
                           [ 8.00909758e-01,  2.38983199e-01,  4.15492684e-01, ...,
                             3.90530713e-02,  2.34373003e-01,  1.22278236e-01],
                           [ 2.60862788e-01,  4.43267114e-02,  3.63648295e-01, ...,
                            -7.53704458e-04,  3.84620279e-02, -2.14213312e-01],
                           [-2.30111778e-01, -4.98388559e-01, -1.26496106e-02, ...,
                             4.49867934e-01,  6.16019145e-02, -2.61357218e-01]]],
                          dtype=float32)>),
                    ('pooler_output',
                    <tf.Tensor: shape=(2, 768), dtype=float32, numpy=
                    array([[-0.9204854 , -0.37138987, -0.6051259 , ..., -0.4473697 ,
                            -0.64347583,  0.9423271 ],
                           [-0.8854158 , -0.26547667,  0.21815054, ...,  0.17237163,
                            -0.6402989 ,  0.8888342 ]], dtype=float32)>)])
```

Figure 3.10 – BERT model output

As can be seen from the result, there are two outputs: one for the last hidden state and one for the pooler output. The last hidden state provides all token embeddings from BERT with additional *[CLS]* and *[SEP]* tokens at the start and end, respectively.

23. Now that you have learned more about the TensorFlow version of BERT, you can make a Keras model using this new embedding:

```
from tensorflow import keras
import tensorflow as tf
max_length = 256
tokens = keras.layers.Input(shape=(max_length,),
                    dtype=tf.dtypes.int32)
masks = keras.layers.Input(shape=(max_length,),
                    dtype=tf.dtypes.int32)
embedding_layer = bert.layers[0]([tokens,masks])[0]
[:,0,:]
```

```
dense = tf.keras.layers.Dense(units=2,
        activation="softmax")(embedding_layer)
model = keras.Model([tokens,masks],dense)
```

24. The model object, which is a Keras model, has two inputs: one for tokens and one for masks. Tokens has token_ids from the tokenizer output and the masks will have attention_mask. Let's try it and see what happens:

```
>>> tokenized = tokenizer.batch_encode_plus(
["hello how is it going with you",
"hello how is it going with you"],
return_tensors="tf",
max_length= max_length,
truncation=True,
pad_to_max_length=True)
```

25. It is important to use max_length, truncation, and pad_to_max_length when using tokenizer. These parameters make sure you have the output in a usable shape by padding it to the maximum length of 256 that was defined before. Now you can run the model using this sample:

```
>>>model([tokenized["input_ids"],tokenized["attention_
mask"]])
```

The following is the output:

```
<tf.Tensor: shape=(2, 2), dtype=float32, numpy=
array([[0.45928752, 0.5407125 ],
       [0.45928752, 0.5407125 ]], dtype=float32)>
```

Figure 3.11 – BERT model classification output

26. When training the model, you need to compile it using the compile function:

```
>>> model.compile(optimizer="Adam",
loss="categorical_crossentropy",
metrics=["accuracy"])
>>> model.summary()
```

The output is as follows:

```
Layer (type)                   Output Shape      Param #    Connected to
==================================================================================
input_tokens (InputLayer)      [(None, 256)]         0

input_masks (InputLayer)       [(None, 256)]         0

bert (TFBertMainLayer)         multiple          109482240  input_tokens[0][0]
                                                            input_masks[0][0]

tf._operators_.getitem_3 (Sli  (None, 768)           0      bert[3][0]

output_layer (Dense)           (None, 2)           1538     tf._operators_.getitem_3[0][0]
==================================================================================
Total params: 109,483,778
Trainable params: 109,483,778
Non-trainable params: 0
```

Figure 3.12 – BERT model summary

27. From the model summary, you can see that the model has 109,483,778 trainable parameters including BERT. But if you have your BERT model pretrained and you want to freeze it in a task-specific training, you can do so with the following command:

```
>>> model.layers[2].trainable = False
```

As far as we know, the layer index of the embedding layer is 2, so we can simply freeze it. If you rerun the summary function, you will see the trainable parameters are reduced to 1,538, which is the number of parameters of the last layer:

```
Layer (type)                   Output Shape      Param #    Connected to
==================================================================================
input_tokens (InputLayer)      [(None, 256)]         0

input_masks (InputLayer)       [(None, 256)]         0

bert (TFBertMainLayer)         multiple          109482240  input_tokens[0][0]
                                                            input_masks[0][0]

tf._operators_.getitem_3 (Sli  (None, 768)           0      bert[3][0]

output_layer (Dense)           (None, 2)           1538     tf._operators_.getitem_3[0][0]
==================================================================================
Total params: 109,483,778
Trainable params: 1,538
Non-trainable params: 109,482,240
```

Figure 3.13 – BERT model summary with fewer trainable parameters

28. As you recall, we used the IMDB sentiment analysis dataset for training the language model. Now you can use it for training the Keras-based model for sentiment analysis. But first, you need to prepare the input and output:

```
import pandas as pd
imdb_df = pd.read_csv("IMDB Dataset.csv")
reviews = list(imdb_df.review)
tokenized_reviews = \
tokenizer.batch_encode_plus(reviews, return_tensors="tf",
                            max_length=max_length,
                            truncation=True,
                            pad_to_max_length=True)
import numpy as np
train_split = int(0.8 * \ len(tokenized_
reviews["attention_mask"]))
train_tokens = tokenized_reviews["input_ids"]\
[:train_split]
test_tokens = tokenized_reviews["input_ids"][train_
split:]
train_masks = tokenized_reviews["attention_mask"]\
[:train_split]
test_masks = tokenized_reviews["attention_mask"]\
[train_split:]
sentiments = list(imdb_df.sentiment)
labels = np.array([[0,1] if sentiment == "positive" else\
[1,0] for sentiment in sentiments])
train_labels = labels[:train_split]
test_labels = labels[train_split:]
```

29. And finally, your data is ready, and you can fit your model:

```
>>> model.fit([train_tokens,train_masks],train_labels,
              epochs=5)
```

And after fitting the model, your model is ready to be used. Up to this point, you have learned how to perform model training for a classification task. You have learned how to save it, and in the next section, you will learn how it is possible to share the trained model with the community.

Sharing models with the community

HuggingFace provides a very easy-to-use model-sharing mechanism:

1. You can simply use the following `cli` tool to log in:

   ```
   Transformers-cli login
   ```

2. After you've logged in using your own credentials, you can create a repository:

   ```
   Transformers-cli repo create a-fancy-model-name
   ```

3. You can put any model name for the `a-fancy-model-name` parameter and then it is essential to make sure you have git-lfs:

   ```
   git lfs install
   ```

   Git LFS is a Git extension used for handling large files. HuggingFace pretrained models are usually large files that require extra libraries such as LFS to be handled by Git.

4. Then you can clone the repository you have just created:

   ```
   git clone https://huggingface.co/username/a-fancy-model-name
   ```

5. Afterward, you can add and remove from the repository as you like, and then, just like Git usage, you have to run the following command:

   ```
   git add . && git commit -m "Update from $USER"
   git push
   ```

Autoencoding models rely on the left encoder side of the original Transformer and are highly efficient at solving classification problems. Even though BERT is a typical example of autoencoding models, there are many alternatives discussed in the literature. Let's take a look at these important alternatives.

Understanding other autoencoding models

In this part, we will review autoencoding model alternatives that slightly modify the original BERT. These alternative re-implementations have led to better downstream tasks by exploiting many sources: optimizing the pre-training process and the number of layers or heads, improving data quality, designing better objective functions, and so forth. The source of improvements roughly falls into two parts: *better architectural design choice* and *pre-training control*.

Many effective alternatives have been shared lately, so it is impossible to understand and explain them all here. We can take a look at some of the most cited models in the literature and the most used ones on NLP benchmarks. Let's start with **Albert** as a re-implementation of BERT that focuses especially on architectural design choice.

Introducing ALBERT

The performance of language models is considered to improve as their size gets bigger. However, training such models is getting more challenging due to both memory limitations and longer training times. To address these issues, the Google team proposed the **Albert model** (**A Lite BERT** for Self-Supervised Learning of Language Representations), which is indeed a reimplementation of the BERT architecture by utilizing several new techniques that reduce memory consumption and increase the training speed. The new design led to the language models scaling much better than the original BERT. Along with 18 times fewer parameters, Albert trains 1.7 times faster than the original BERT-large model.

The Albert model mainly consists of three modifications of the original BERT:

- Factorized embedding parameterization
- Cross-layer parameter sharing
- Inter-sentence coherence loss

The first two modifications are parameter-reduction methods that are related to the issue of model size and memory consumption in the original BERT. The third corresponds to a new objective function: **Sentence-Order Prediction** (**SOP**), replacing the **Next Sentence Prediction** (**NSP**) task of the original BERT, which led to a much thinner model and improved performance.

Factorized embedding parameterization is used to decompose the large vocabulary-embedding matrix into two small matrices, which separate the size of the hidden layers from the size of the vocabulary. This decomposition reduces the embedding parameters from $O(V \times H)$ to $O(V \times E + E \times H)$ where V is *Vocabulary*, H is *Hidden Layer Size*, E is *Embedings*, which leads to more efficient usage of the total model parameters *if $H >> E$* is satisfied.

Cross-layer parameter sharing prevents the total number of parameters from increasing as the network gets deeper. The technique is considered another way to improve parameter efficiency since we can keep the parameter size smaller by sharing or copying. In the original paper, they experimented with many ways to share parameters, such as either sharing FF-only parameters across layers or sharing attention-only parameters or entire parameters.

The other modification of Albert is inter-sentence coherence loss. As we already discussed, the BERT architecture takes advantage of two loss calculations, the **Masked Language Modeling** (**MLM**) loss and NSP. NSP comes with binary cross-entropy loss for predicting whether or not two segments appear in a row in the original text. The negative examples are obtained by selecting two segments from different documents. However, the Albert team criticized NSP for being a topic detection problem, which is considered a relatively easy problem. Therefore, the team proposed a loss based primarily on coherence rather than topic prediction. They utilized SOP loss, which focuses on modeling inter-sentence coherence instead of topic prediction. SOP loss uses the same positive examples technique as BERT, (which is two consecutive segments from the same document), and as negative examples, the same two consecutive segments but with their order swapped. The model is then forced to learn finer-grained distinctions between coherence properties at the discourse level.

1. Let's compare the original BERT and Albert configuration with the
    Transformers library. The following piece of code shows how to configure a
    BERT-Base initial model. As you see in the output, the number of parameters is
    around 110 M:

    ```
    #BERT-BASE (L=12, H=768, A=12, Total Parameters=110M)
    >> from Transformers import BertConfig, BertModel
    >> bert_base= BertConfig()
    >> model = BertModel(bert_base)
    >> print(f"{model.num_parameters() /(10**6)}\
     million parameters")
    109.48224 million parameters
    ```

2. And the following piece of code shows how to define the Albert model with two
    classes, AlbertConfig and AlbertModel, from the Transformers library:

    ```
    # Albert-base Configuration
    >>> from Transformers import AlbertConfig, AlbertModel
    >>> albert_base = AlbertConfig(hidden_size=768,
                                   num_attention_heads=12,
                                   intermediate_size=3072,)
    >>> model = AlbertModel(albert_base)
    >>> print(f"{model.num_parameters() /(10**6)}\
    million parameters")
    11.683584 million parameters
    ```

Due to that, the default Albert configuration points to Albert-xxlarge. We need to set the hidden size, the number of attention heads, and the intermediate size to fit Albert-base. And the code shows the Albert-base mode as 11M, 10 times smaller than the BERT-base model. The original paper on ALBERT reported benchmarking as in the following table:

Model		Parameters	SQuAD1.1	SQuAD2.0	MNLI	SST-2	RACE	Avg	Speedup
BERT	base	108M	90.4/83.2	80.4/77.6	84.5	92.8	68.2	82.3	4.7x
	large	334M	92.2/85.5	85.0/82.2	86.6	93.0	73.9	85.2	1.0
ALBERT	base	12M	89.3/82.3	80.0/77.1	81.6	90.3	64.0	80.1	5.6x
	large	18M	90.6/83.9	82.3/79.4	83.5	91.7	68.5	82.4	1.7x
	xlarge	60M	92.5/86.1	86.1/83.1	86.4	92.4	74.8	85.5	0.6x
	xxlarge	235M	**94.1/88.3**	**88.1/85.1**	**88.0**	**95.2**	82.3	**88.7**	0.3x

Figure 3.14 – Albert model benchmarking

3. From this point on, in order to train an Albert language model from scratch, we need to go through similar phases to those we already illustrated in BERT training in the previous sections by using the uniform Transformers API. There's no need to explain the same steps here! Instead, let's load an already trained Albert language model as follows:

```
from Transformers import AlbertTokenizer, AlbertModel
tokenizer = \
AlbertTokenizer.from_pretrained("albert-base-v2")
model = AlbertModel.from_pretrained("albert-base-v2")
text = "The cat is so sad ."
encoded_input = tokenizer(text, return_tensors='pt')
output = model(**encoded_input)
```

4. The preceding pieces of code download the Albert model weights and its configuration from the HuggingFace hub or from our local cache directory if already cached, which means you've already called the `AlbertTokenizer.from_pretrained()` function before. Since that the model object is a pre-trained language model, the things we can do with this model are limited for now. We need to train it on a downstream task to able to use it for inference, which will be the main subject of further chapters. Instead, we can take advantage of its masked language model objective as follows:

```
from Transformers import pipeline
fillmask= pipeline('fill-mask', model='albert-base-v2')
pd.DataFrame(fillmask("The cat is so [MASK] ."))
```

The following is the output:

sequence	score	token	token_str
[CLS] the cat is so cute.[SEP]	0.281025	10901	_cute
[CLS] the cat is so adorable.[SEP]	0.094893	26354	_adorable
[CLS] the cat is so happy.[SEP]	0.042963	1700	_happy
[CLS] the cat is so funny.[SEP]	0.040976	5066	_funny
[CLS] the cat is so affectionate.[SEP]	0.024233	28803	_affectionate

Figure 3.15 – The fill-mask output results for albert-base-v2

The `fill-mask` pipeline computes the scores for each vocabulary token with the `SoftMax()` function and sorts the most probable tokens where `cute` is the winner with a probability score of 0.281. You may notice that entries in the *token_str* column start with the _ character, which is due to the metaspace component of the tokenizer of Albert.

Let's take a look at the next alternative, *RoBERTa*, which mostly focuses on the pre-training phase.

RoBERTa

Robustly Optimized BERT pre-training Approach (**RoBERTa**) is another popular BERT reimplementation. It has provided many more improvements in training strategy than architectural design. It outperformed BERT in almost all individual tasks on GLUE. Dynamic masking is one of its original design choices. Although static masking is better for some tasks, the RoBERTa team showed that dynamic masking can perform well for overall performances. Let's compare the changes from BERT and summarize all the features as follows:

The changes in architecture are as follows:

- Removing the next sentence prediction training objective
- Dynamically changing the masking patterns instead of static masking, which is done by generating masking patterns whenever they feed a sequence to the model
- **BPE** sub-word tokenizer

The changes in training are as follows:

- Controlling the training data: More data is used, such as 160 GB instead of the 16 GB originally used in BERT. Not only the size of the data but the quality and diversity were taken into consideration in the study.

- Longer iterations of up to 500K pretraining steps.

- A longer batch size.

- Longer sequences, which leads to less padding.

- A large 50K BPE vocabulary instead of a 30K BPE vocabulary.

Thanks to the Transformers uniform API, as in the Albert model pipeline above, we initialize the RoBERTa model as follows:

```
>>> from Transformers import RobertaConfig, RobertaModel
>>> conf= RobertaConfig()
>>> model = RobertaModel(conf)
>>> print(f"{model.num_parameters() /(10**6)}\
million parameters")
109.48224 million parameters
```

In order to load the pre-trained model, we execute the following pieces of code:

```
from Transformers import RobertaTokenizer, RobertaModel
tokenizer = \
RobertaTokenizer.from_pretrained('roberta-base')
model = RobertaModel.from_pretrained('roberta-base')
text = "The cat is so sad ."
encoded_input = tokenizer(text, return_tensors='pt')
output = model(**encoded_input)
```

These lines illustrate how the model processes a given text. The output representation at the last layer is not useful at the moment. As we've mentioned several times, we need to fine-tune the main language models. The following execution applies the `fill-mask` function using the `roberta-base` model:

```
>>> from Transformers import pipeline
>>> fillmask= pipeline("fill-mask ",model="roberta-base",
                        tokenizer=tokenizer)
>>> pd.DataFrame(fillmask("The cat is so <mask> ."))
```

The following is the output:

sequence	score	token	token_str
<s>The cat is so cute.</s>	0.191843	11962	Ġcute
<s>The cat is so sweet.</s>	0.051524	4045	Ġsweet
<s>The cat is so funny.</s>	0.033595	6269	Ġfunny
<s>The cat is so handsome.</s>	0.032893	19222	Ġhandsome
<s>The cat is so beautiful.</s>	0.032314	2721	Ġbeautiful

Figure 3.16 – The fill-mask task results for roberta-base

Like the previous ALBERT `fill-mask` model, this pipeline ranks the suitable candidate words. Please ignore the prefix Ġ in the tokens – that is an encoded space character produced by the byte-level BPE tokenizer, which we will discuss later. You should have noticed that we used the `[MASK]` and `<mask>` tokens in ALBERT and RoBERTa pipeline in order to hold place for masked token. This is because of the configuration of `tokenizer`. To learn which token expression will be used, you can check `tokenizer.mask_token`. Please see the following execution:

```
>>> tokenizer = \
  AlbertTokenizer.from_pretrained('albert-base-v2')
>>> print(tokenizer.mask_token)
[MASK]
>>> tokenizer = \
RobertaTokenizer.from_pretrained('roberta-base')
>>> print(tokenizer.mask_token)
<mask>
```

To ensure proper mask token use, we can add the `fillmask.tokenizer.mask_token` expression in the pipeline as follows:

```
fillmask(f"The cat is very\
{fillmask.tokenizer.mask_token}.")
```

ELECTRA

The **ELECTRA** model (*proposed by Kevin Clark et al. in 2020*) focuses on a new masked language model utilizing the replaced token detection training objective. During pre-training, the model is forced to learn to distinguish real input tokens from synthetically generated replacements where the synthetic negative example is sampled from plausible tokens rather than randomly sampled tokens. The Albert model criticized the NSP objective of BERT for being a topic detection problem and using low-quality negative examples. ELECTRA trains two neural networks, a generator and a discriminator, so that the former produces high-quality negative examples, whereas the latter distinguishes the original token from the replaced token. We know GAN networks from the field of computer vision, in which the generator *G* produces fake images and tries to fool the discriminator *D*, and the discriminator network tries to avoid being fooled. The ELECTRA model applies almost the same generator-discriminator approach to replace original tokens with high-quality negative examples that are plausible replacements but synthetically generated.

In order not to repeat the same code with other examples, we only provide a simple `fill-mask` example for the Electra generator as follows:

```
fillmask = \
pipeline("fill-mask", model="google/electra-small-generator")
fillmask(f"The cat is very \{fillmask.tokenizer.mask_token} .")
```

You can see the entire list of models at the following link: `https://huggingface.co/Transformers/model_summary.html`.

The model checkpoints can be found at `https://huggingface.co/models`.

Well done! We've finally completed the autoencoding model part. Now we'll move on to tokenization algorithms, which have an important effect on the success of Transformers.

Working with tokenization algorithms

In the opening part of the chapter, we trained the BERT model using a specific tokenizer, namely `BertWordPieceTokenizer`. Now it is worth discussing the tokenization process in detail here. Tokenization is a way of splitting textual input into tokens and assigning an identifier to each token before feeding the neural network architecture. The most intuitive way is to split the sequence into smaller chunks in terms of space. However, such approaches do not meet the requirement of some languages, such as Japanese, and also may lead to huge vocabulary problems. Almost all Transformer models leverage subword tokenization not only for reducing dimensionality but also for encoding rare (or unknown) words not seen in training. The tokenization relies on the idea that every word, including rare words or unknown words, can be decomposed into meaningful smaller chunks that are widely seen symbols in the training corpus.

Some traditional tokenizers developed within Moses and the `nltk` library apply advanced rule-based techniques. But the tokenization algorithms that are used with Transformers are based on self-supervised learning and extract the rules from the corpus. Simple intuitive solutions for rule-based tokenization are based on using characters, punctuation, or whitespace. Character-based tokenization causes language models to lose the input meaning. Even though it can reduce the vocabulary size, which is good, it makes it hard for the model to capture the meaning of `cat` by means of the encodings of the characters `c`, `a`, and `t`. Moreover, the dimension of the input sequence becomes very large. Likewise, punctuation-based models cannot treat some expressions, such as *haven't* or *ain't*, properly.

Recently, several advanced subword tokenization algorithms, such as BPE, have become an integral part of Transformer architectures. These modern tokenization procedures consist of two phases: The pre-tokenization phase simply splits the input into tokens either using space as or language-dependent rules. Second, the tokenization training phase is to train the tokenizer and build a base vocabulary of a reasonable size based on tokens. Before training our own tokenizers, let's load a pre-trained tokenizer. The following code loads a Turkish tokenizer, which is of type `BertTokenizerFast`, from the `Transformers` library with a vocabulary size of 32K:

```
>>> from Transformers import AutoModel, AutoTokenizer
>>> tokenizerTUR = AutoTokenizer.from_pretrained(
                    "dbmdz/bert-base-turkish-uncased")
>>> print(f"VOC size is: {tokenizerTUR.vocab_size}")
>>> print(f"The model is: {type(tokenizerTUR)}")
VOC size is: 32000
The model is: Transformers.models.bert.tokenization_bert_fast.
BertTokenizerFast
```

The following code loads an English BERT tokenizer for the `bert-base-uncased` model:

```
>>> from Transformers import AutoModel, AutoTokenizer
>>> tokenizerEN = \
 AutoTokenizer.from_pretrained("bert-base-uncased")
>>> print(f"VOC size is: {tokenizerEN.vocab_size}")
>>> print(f"The model is {type(tokenizerEN)}")
VOC size is: 30522
The model is ... BertTokenizerFast
```

Let's see how they work! We tokenize the word `telecommunication` with these two tokenizers:

```
>>> word_en="telecommunication"
>>> print(f"is in Turkish Model ? \
{word_en in tokenizerTUR.vocab}")
>>> print(f"is in English Model ? \
{word_en in tokenizerEN.vocab}")
is in Turkish Model ? False
is in English Model ? True
```

The `word_en` token is already in the vocabulary of the English tokenizer but not in that of the Turkish one. So let's see what happens with the Turkish tokenizer:

```
>>> tokens=tokenizerTUR.tokenize(word_en)
>>> tokens
['tel', '##eco', '##mm', '##un', '##ica', '##tion']
```

Since the Turkish tokenizer model has no such a word in its vocabulary, it needs to break the word into parts that make sense to it. All these split tokens are already stored in the model vocabulary. Please notice the output of the following execution:

```
>>> [t in tokenizerTUR.vocab for t in tokens]
[True, True, True, True, True, True]
```

Let's tokenize the same word with the English tokenizer that we already loaded:

```
>>> tokenizerEN.tokenize(word_en)
['telecommunication']
```

Since the English model has the word `telecommunication` in the base vocabulary, it does not need to break it into parts but rather takes it as a whole. By learning from the corpus, the tokenizers are capable of transforming a word into mostly grammatically logical subcomponents. Let's take a difficult example from Turkish. As an agglutinative language, Turkish allows us to add many suffixes to a word stem to construct very long words. Here is one of the longest words in the Turkish language used in a text (`https://en.wikipedia.org/wiki/Longest_word_in_Turkish`):

Muvaffakiyetsizleştiricileştiriveremeyebileceklerimizdenmişsinizcesine

It means that *As though you happen to have been from among those whom we will not be able to easily/quickly make a maker of unsuccessful ones.* The Turkish BERT tokenizer may not have seen this word in training, but it has seen its pieces; *muvaffak (succesful) as the stem, ##iyet(successfulness), ##siz (unsuccessfulness), ##leş (become unsuccessful),* and so forth. The Turkish tokenizer extracts components that seem to be grammatically logical for the Turkish language when comparing the results with a Wikipedia article:

```
>>> print(tokenizerTUR.tokenize(long_word_tur))
['muvaffak', '##iyet', '##siz', '##les', '##tir', '##ici',
 '##les', '##tir', '##iver', '##emeye', '##bilecekleri', '##mi',
 '##z', '##den', '##mis', '##siniz', '##cesine']
```

The Turkish tokenizer is an example of the WordPiece algorithm since it works with a BERT model. Almost all language models including BERT, DistilBERT, and ELECTRA require a WordPiece tokenizer.

Now we are ready to take a look at the tokenization approaches used with Transformers. First, we'll discuss the widely used tokenizations of BPE, WordPiece, and SentencePiece a bit and then train them with HuggingFace's fast `tokenizers` library.

Byte pair encoding

BPE is a data compression technique. It scans the data sequence and iteratively replaces the most frequent pair of bytes with a single symbol. It was first adapted and proposed in *Neural Machine Translation of Rare Words with Subword Units, Sennrich et al. 2015,* to solve the problem of unknown words and rare words for machine translation. Currently, it is successfully being used within GPT-2 and many other state-of-the-art models. Many modern tokenization algorithms are based on such compression techniques.

It represents text as a sequence of character n-grams, which are also called character-level subwords. The training starts initially with a vocabulary of all Unicode characters (or symbols) seen in the corpus. This can be small for English but can be large for character-rich languages such as Japanese. Then, it iteratively computes character bigrams and replaces the most frequent ones with special new symbols. For example, *t* and *h* are frequently occurring symbols. We replace consecutive symbols with the *th* symbol. This process is kept iteratively running until the vocabulary has attained the desired vocabulary size. The most common vocabulary size is around 30K.

BPE is particularly effective at representing unknown words. However, it may not guarantee the handling of rare words and/or words including rare subwords. In such cases, it associates rare characters with a special symbol, *<UNK>*, which may lead to losing meaning in words a bit. As a potential solution, **Byte-Level BPE** (**BBPE**) has been proposed, which uses a 256-byte set of vocabulary instead of Unicode characters to ensure that every base character is included in the vocabulary.

WordPiece tokenization

WordPiece is another popular word segmentation algorithm widely used with BERT, DistilBERT, and Electra. It was proposed by Schuster and Nakajima to solve the Japanese and Korean voice problem in 2012. The motivation behind this work was that, although not a big issue for the English language, word segmentation is important preprocessing for many Asian languages, because in these languages spaces are rarely used. Therefore, we come across word segmentation approaches in NLP studies in Asian languages more often. Similar to BPE, WordPiece uses a large corpus to learn vocabulary and merging rules. While BPE and BBPE learn the merging rules based on co-occurrence statistics, the WordPiece algorithm uses maximum likelihood estimation to extract the merging rules from a corpus. It first initializes the vocabulary with Unicode characters, which are also called vocabulary symbols. It treats each word in the training corpus as a list of symbols (initially Unicode characters), and then it iteratively produces a new symbol merging two symbols out of all the possible candidate symbol pairs based on the likelihood maximization rather than frequency. This production pipeline continues until the desired vocabulary size is reached.

Sentence piece tokenization

Previous tokenization algorithms treat text as a space-separated word list. This space-based splitting does not work in some languages. In the German language, compound nouns are written without spaces, for example, menschenrechte (human rights). The solution is to use language-specific pre-tokenizers. In German, an NLP pipeline leverages a compound-splitter module to check whether a word can be subdivided into smaller words. However, East Asian languages (for example, Chinese, Japanese, Korean, and Thai) do not use spaces between words. The **SentencePiece** algorithm is designed to overcome this space limitation, which is a simple and language-independent tokenizer proposed by Kudo et al. in 2018. It treats the input as a raw input stream where space is part of the character set. The tokenizer using SentencePiece produces the _ character, which is also why we saw _ in the output of the Albert model example earlier. Other popular language models that use SentencePiece are XLNet, Marian, and T5.

So far, we have discussed subword tokenization approaches. It is time to start conducting experiments for training with the `tokenizers` library.

The tokenizers library

You may have noticed that the already-trained tokenizers for Turkish and English are part of the `Transformers` library in the previous code examples. On the other hand, the HuggingFace team provided the `tokenizers` library independently from the `Transformers` library to be fast and give us more freedom. The library was originally written in Rust, which makes multi-core parallel computations possible and is wrapped with Python (`https://github.com/huggingface/tokenizers`).

To install the `tokenizers` library, we use this:

```
$ pip install tokenizers
```

The `tokenizers` library provides several components so that we can build an end-to-end tokenizer from preprocessing the raw text to decoding tokenized unit IDs:

Normalizer→ PreTokenizer → Modeling → Post-Processor → Decoding

The following diagram depicts the tokenization pipeline:

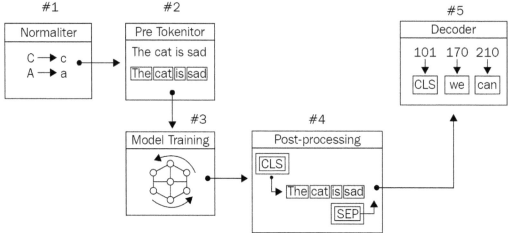

Figure 3.17 – Tokenization pipeline

- **Normalizer** allows us to apply primitive text processing such as lowercasing, stripping, Unicode normalization, and removing accents.

- **PreTokenizer** prepares the corpus for the next training phase. It splits the input into tokens depending on the rules, such as whitespace.

- **Model Training** is a subword tokenization algorithm such as *BPE*, *BBPE*, and *WordPiece*, which we've discussed already. It discovers subwords/vocabulary and learns generation rules.

- **Post-processing** provides advanced class construction that is compatible with Transformers models such as BertProcessors. We mostly add special tokens such as *[CLS]* and *[SEP]* to the tokenized input just before feeding the architecture.

- **Decoder** is in charge of converting token IDs back to the original string. It is just for inspecting what is going on.

Training BPE

Let's train a BPE tokenizer using Shakespeare's plays:

1. It is loaded as follows:

```
import nltk
from nltk.corpus import gutenberg
nltk.download('gutenberg')
nltk.download('punkt')
```

```
plays=['shakespeare-macbeth.txt','shakespeare-hamlet.
txt',
       'shakespeare-caesar.txt']
shakespeare=[" ".join(s) for ply in plays \
for s in gutenberg.sents(ply)]
```

We will need a post-processor (`TemplateProcessing`) for all the tokenization algorithms ahead. We need to customize the post-processor to make the input convenient for a particular language model. For example, the following template will be suitable for the BERT model since it needs the *[CLS]* token at the beginning of the input and *[SEP]* tokens both at the end and in the middle.

2. We define the template as follows:

```
from tokenizers.processors import TemplateProcessing
special_tokens=["[UNK]","[CLS]","[SEP]","[PAD]","[MASK]"]
temp_proc= TemplateProcessing(
    single="[CLS] $A [SEP]",
    pair="[CLS] $A [SEP] $B:1 [SEP]:1",
    special_tokens=[
        ("[CLS]", special_tokens.index("[CLS]")),
        ("[SEP]", special_tokens.index("[SEP]")),
    ],
)
```

3. We import the necessary components to build an end-to-end tokenization pipeline:

```
from tokenizers import Tokenizer
from tokenizers.normalizers import \
(Sequence,Lowercase, NFD, StripAccents)
from tokenizers.pre_tokenizers import Whitespace
from tokenizers.models import BPE
from tokenizers.decoders import BPEDecoder
```

4. We start by instantiating **BPE** as follows:

```
tokenizer = Tokenizer(BPE())
```

5. The preprocessing part has two components: *normalizer* and *pre-tokenizer*. We may have more than one normalizer. So, we compose a `Sequence` of normalizer components that includes multiple normalizers where `NFD()` is a Unicode normalizer and `StripAccents()` removes accents. For pre-tokenization, `Whitespace()` gently breaks the text based on space. Since the decoder component must be compatible with the model, `BPEDecoder` is selected for the BPE model:

```
tokenizer.normalizer = Sequence(
[NFD(),Lowercase(),StripAccents()])
tokenizer.pre_tokenizer = Whitespace()
tokenizer.decoder = BPEDecoder()
tokenizer.post_processor=temp_proc
```

6. Well! We are ready to train the tokenizer on the data. The following execution instantiates `BpeTrainer()`, which helps us to organize the entire training process by setting hyperparameters. We set the vocabulary size parameter to 5K since our Shakespeare corpus is relatively small. For a large-scale project, we use a bigger corpus and normally set the vocabulary size to around 30K:

```
>>> from tokenizers.trainers import BpeTrainer
>>> trainer = BpeTrainer(vocab_size=5000,
                        special_tokens= special_tokens)
>>> tokenizer.train_from_iterator(shakespeare,
                        trainer=trainer)
>>> print(f"Trained vocab size:\
{tokenizer.get_vocab_size()}" )
Trained vocab size: 5000
```

We have completed the training!

> **Important note**
> Training from the filesystem: To start the training process, we passed an in-memory Shakespeare object as a list of strings to `tokenizer.train_from_iterator()`. For a large-scale project with a large corpus, we need to design a Python generator that yields string lines mostly by consuming the files from the filesystem rather than in-memory storage. You should also check `tokenizer.train()` to train from the filesystem storage as applied in the BERT training section above.

7. Let's grab a random sentence from the play Macbeth, name it `sen`, and tokenize it with our fresh tokenizer:

```
>>> sen= "Is this a dagger which I see before me,\
 the handle toward my hand?"
>>> sen_enc=tokenizer.encode(sen)
>>> print(f"Output: {format(sen_enc.tokens)}")
Output: ['[CLS]', 'is', 'this', 'a', 'dagger', 'which',
'i', 'see', 'before', 'me', ',', 'the', 'hand', 'le',
'toward', 'my', 'hand', '?', '[SEP]']
```

8. Thanks to the post-processor function above, we see additional *[CLS]* and *[SEP]* tokens in the proper position. There is only one split word, *handle* (*hand, le*), since we passed to the model a sentence from the play Macbeth that the model already knew. Besides, we used a small corpus, and the tokenizer is not forced to use compression. Let's pass a challenging phrase, `Hugging Face`, that the tokenizer might not know:

```
>>> sen_enc2=tokenizer.encode("Macbeth and Hugging Face")
>>> print(f"Output: {format(sen_enc2.tokens)}")
Output: ['[CLS]', 'macbeth', 'and', 'hu', 'gg', 'ing',
'face', '[SEP]']
```

9. The term `Hugging` is lowercased and split into three pieces `hu`, `gg`, `ing`, since the model vocabulary contains all other tokens but `Hugging`. Let's pass two sentences now:

```
>>> two_enc=tokenizer.encode("I like Hugging Face!",
"He likes Macbeth!")
>>> print(f"Output: {format(two_enc.tokens)}")
Output: ['[CLS]', 'i', 'like', 'hu', 'gg', 'ing', 'face',
'!', '[SEP]', 'he', 'li', 'kes', 'macbeth', '!', '[SEP]']
```

Notice that the post-processor injected the `[SEP]` token as an indicator.

10. It is time to save the model. We can either save the sub-word tokenization model or the entire tokenization pipeline. First, let's save the BPE model only:

```
>>> tokenizer.model.save('.')
['./vocab.json', './merges.txt']
```

11. The model saved two files regarding vocabulary and merging rules. The `merge.` `txt` file is composed of 4,948 merging rules:

```
$ wc -l ./merges.txt
4948 ./merges.txt
```

12. The top five rules ranked are as shown in the following where we see that (t, h) is the first ranked rule due to that being the most frequent pair. For testing, the model scans the textual input and tries to merge these two symbols first if applicable:

```
$ head -3 ./merges.txt
t h
o u
a n
th e
r e
```

The BPE algorithm ranks the rules based on frequency. When you manually calculate character bigrams in the Shakespeare corpus, you will find (t, h) the most frequent pair.

13. Let's now save and load the entire tokenization pipeline:

```
>>> tokenizer.save("MyBPETokenizer.json")
>>> tokenizerFromFile = \
Tokenizer.from_file("MyBPETokenizer.json")
>>> sen_enc3 = \
tokenizerFromFile.encode("I like Hugging Face and
Macbeth")
>>> print(f"Output: {format(sen_enc3.tokens)}")
Output: ['[CLS]', 'i', 'like', 'hu', 'gg', 'ing', 'face',
'and', 'macbeth', '[SEP]']
```

We successfully reloaded the tokenizer!

Training the WordPiece model

In this section, we will train the WordPiece model:

1. We start by importing the necessary modules:

    ```
    from tokenizers.models import WordPiece
    from tokenizers.decoders import WordPiece \
    as WordPieceDecoder
    from tokenizers.normalizers import BertNormalizer
    ```

2. The following lines instantiate an empty WordPiece tokenizer and prepare it for training. `BertNormalizer` is a pre-defined normalizer sequence that includes the processes of cleaning the text, transforming accents, handling Chinese characters, and lowercasing:

    ```
    tokenizer = Tokenizer(WordPiece())
    tokenizer.normalizer=BertNormalizer()
    tokenizer.pre_tokenizer = Whitespace()
    tokenizer.decoder= WordPieceDecoder()
    ```

3. Now, we instantiate a proper trainer, `WordPieceTrainer()` for `WordPiece()`, to organize the training process:

    ```
    >>> from tokenizers.trainers import WordPieceTrainer
    >>> trainer = WordPieceTrainer(vocab_size=5000,\
                special_tokens=["[UNK]", "[CLS]", "[SEP]",\
                "[PAD]", "[MASK]"])
    >>> tokenizer.train_from_iterator(shakespeare,
    trainer=trainer)
    >>> output = tokenizer.encode(sen)
    >>> print(output.tokens)
    ['is', 'this', 'a', 'dagger', 'which', 'i', 'see',
    'before', 'me', ',', 'the', 'hand', '##le', 'toward',
    'my', 'hand', '?']
    ```

4. Let's use `WordPieceDecoder()` to treat the sentences properly:

    ```
    >>> tokenizer.decode(output.ids)
    'is this a dagger which i see before me, the handle
    toward my hand?'
    ```

5. We have not come across any [UNK] tokens in the output since the tokenizer somehow knows or splits the input for encoding. Let's force the model to produce [UNK] tokens as in the following code. Let's pass a Turkish sentence to our tokenizer:

```
>>> tokenizer.encode("Kralsın aslansın Macbeth!").tokens
'[UNK]', '[UNK]', 'macbeth', '!']
```

Well done! We have a couple of unknown tokens since the tokenizer does not find a way to decompose the given word from the merging rules and the base vocabulary.

So far, we have designed our tokenization pipeline all the way from the normalizer component to the decoder component. On the other hand, the tokenizers library provides us with an already made (not trained) empty tokenization pipeline with proper components to build quick prototypes for production. Here are some pre-made tokenizers:

- CharBPETokenizer: The original BPE
- ByteLevelBPETokenizer: The byte-level version of the BPE
- SentencePieceBPETokenizer: A BPE implementation compatible with the one used by *SentencePiece*
- BertWordPieceTokenizer: The famous BERT tokenizer, using WordPiece

The following code imports these pipelines:

```
>>> from tokenizers import (ByteLevelBPETokenizer,
                            CharBPETokenizer,
                            SentencePieceBPETokenizer,
                            BertWordPieceTokenizer)
```

All these pipelines are already designed for us. The rest of the process (such as training, saving the model, and using the tokenizer) is the same as our previous BPE and WordPiece training procedure.

Well done! We have made great progress and trained our first Transformer model as well as its tokenizer.

Summary

In this chapter, we have experienced autoencoding models both theoretically and practically. Starting with basic knowledge about BERT, we trained it as well as a corresponding tokenizer from scratch. We also discussed how to work inside other frameworks, such as Keras. Besides BERT, we also reviewed other autoencoding models. To avoid excessive code repetition, we did not provide the full implementation for training other models. During the BERT training, we trained the WordPiece tokenization algorithm. In the last part, we examined other tokenization algorithms since it is worth discussing and understanding all of them.

Autoencoding models use the left decoder side of the original Transformer and are mostly fine-tuned for classification problems. In the next chapter, we will discuss and learn about the right decoder part of Transformers to implement language generation models.

4
Autoregressive and Other Language Models

We looked at details of **Autoencoder** (**AE**) language models in *Chapter 3, Autoencoding Language Models*, and studied how an AE language model can be trained from scratch. In the current chapter, you will see theoretical details of **Autoregressive** (**AR**) language models and learn how to pre-train them on your own corpus. You will learn how to pre-train any language model such as **Generated Pre-trained Transformer 2** (**GPT-2**) on your own text and use it in various tasks such as **Natural Language Generation** (**NLG**). You will understand the basics of a **Text-to-Text Transfer Transformer** (**T5**) model and train a **Multilingual T5** (**mT5**) model on your own **Machine Translation** (**MT**) data. After finishing this chapter, you will have an overview of AR language models and their various use cases in text2text applications, such as summarization, paraphrasing, and MT.

The following topics will be covered in this chapter:

- Working with AR language models
- Working with **Sequence-to-Sequence** (**Seq2Seq**) models
- AR language model training

- NLG using AR models
- Summarization and MT fine-tuning using `simpletransformers`

Technical requirements

The following libraries/packages are required to successfully complete this chapter:

- Anaconda
- `transformers 4.0.0`
- `pytorch 1.0.2`
- `tensorflow 2.4.0`
- `datasets 1.4.1`
- `tokenizers`
- `simpletransformers 0.61`

All notebooks with coding exercises will be available at the following GitHub link: `https://github.com/PacktPublishing/Mastering-Transformers/tree/main/CH04`.

Check out the following link to see the Code in Action: `https://bit.ly/3yjn55X`

Working with AR language models

The Transformer architecture was originally intended to be effective for Seq2Seq tasks such as MT or summarization, but it has since been used in diverse NLP problems ranging from token classification to coreference resolution. Subsequent works began to use separately and more creatively the left and right parts of the architecture. The objective, also known as **denoising objective**, is to fully recover the original input from the corrupted one in a bidirectional fashion, as shown on the left side of *Figure 4.1*, which you will see shortly. As seen in the **Bidirectional Encoder Representations from Transformers (BERT)** architecture, which is a notable example of AE models, they can incorporate the context of both sides of a word. However, the first issue is that the corrupting [MASK] symbols that are used during the pre-training phase are absent from the data during the fine-tuning phase, leading to a pre-training-fine-tuning discrepancy. Secondly, the BERT model arguably assumes that the masked tokens are independent of each other.

On the other hand, AR models keep away from such assumptions regarding independence and do not naturally suffer from the pre-train-fine-tuning discrepancy because they rely on the objective predicting the next token conditioned on the previous tokens without masking them. They merely utilize the decoder part of the transformer with masked self-attention. They prevent the model from accessing words to the right of the current word in a forward direction (or to the left of the current word in a backward direction), which is called **unidirectionality**. They are also called **Causal Language Models** (**CLMs**) due to their unidirectionality.

The difference between AE and AR models is simply depicted here:

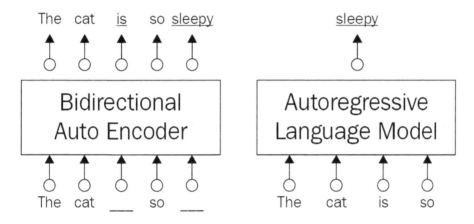

Figure 4.1 – AE versus AR language model

GPT and its two successors (GPT-2, GPT-3), **Transformer-XL**, and **XLNet** are among the popular AR models in the literature. Even though XLNet is based on autoregression, it somehow managed to make use of both contextual sides of the word in a bidirectional fashion, with the help of the permutation-based language objective. Now, we start introducing them and show how to train the models with a variety of experiments. Let's look at GPTs first.

Introduction and training models with GPT

AR models are made up of multiple transformer blocks. Each block contains a masked multi-head self-attention layer along with a pointwise feed-forward layer. The activation in the final transformer block is fed into a softmax function that produces the word-probability distributions over an entire vocabulary of words to predict the next word.

In the original GPT paper *Improving Language Understanding by Generative Pre-Training* (2018), the authors addressed several bottlenecks that traditional **Machine Learning (ML)**-based **Natural Language Processing (NLP)** pipelines are subject to. For example, these pipelines firstly require both a massive amount of task-specific data and task-specific architecture. Secondly, it is hard to apply task-aware input transformations with minimal changes to the architecture of the pre-trained model. The original GPT and its successors (GPT-2 and GPT-3), designed by the OpenAI team, have focused on solutions to alleviate these bottlenecks. The major contribution of the original GPT study is that the pre-trained model achieved satisfactory results, not only for a single task but a diversity of tasks. Having learned the generative model from unlabeled data, which is called unsupervised pre-training, the model is simply fine-tuned to a downstream task by a relatively small amount of task-specific data, which is called **supervised fine-tuning**. This two-stage scheme is widely used in other transformer models, where unsupervised pre-training is followed by supervised fine-tuning.

To keep the GPT architecture as generic as possible, only the inputs are transformed into a task-specific manner, while the entire architecture is kept almost the same. This traversal-style approach converts the textual input into an ordered sequence according to the task so that the pre-trained model can understand the task from it. The left part of *Figure 4.2* (inspired from the original paper) illustrates the transformer architecture and training objectives used in the original GPT work. The right part shows how to transform input for fine-tuning on several tasks.

To put it simply, for a single-sequence task such as text classification, the input is passed through the network as-is, and the linear layer takes the last activations to make a decision. For sentence-pair tasks such as textual entailment, the input that is made up of two sequences is marked with a delimiter, shown as the second example in *Figure 4.2*. In both scenarios, the architecture sees uniform token sequences be processed by the pre-trained model. The delimiter used in this transformation helps the pre-trained model to know which part is premise or hypothesis in the case of textual entailment. Thanks to input transformation, we do not have to make substantial changes in the architecture across the tasks.

You can see a representation of input transformation here:

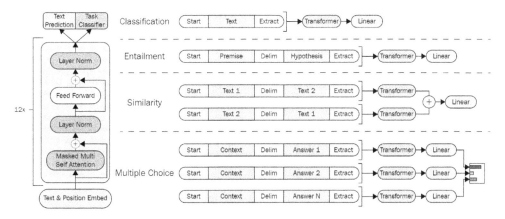

Figure 4.2 – Input transformation (inspired from the paper)

The GPT and its two successors mostly focused on seeking a particular architectural design where the fine-tuning phase was not required. It is based on the idea that a model can be very skilled in the sense that it can learn much of the information about a language during the pre-training phase, with little work left for the fine-tuning phase. Thus, the fine-tuning process can be completed within three epochs and with relatively small examples for most of the tasks. In an extreme case, zero-shot learning aims to disable the fine-tuning phase. The underlying idea is that the model can learn much information about the language during pre-training. This is especially true for all transformer-based models.

Successors of the original GPTs

GPT-2 (see the paper *Language Models are Unsupervised Multitask Learners* (2019)), a successor to the original GPT-1, is a larger model trained on much more training data, called WebText, than the original one. It achieved state-of-the-art results on seven out of the eight tasks in a zero-shot setting in which there is no fine-tuning applied but had limited success in some tasks. It achieved comparable results on smaller datasets for measuring long-range dependency. The GPT-2 authors argued that language models do not necessarily need explicit supervision to learn a task. Instead, they can learn these tasks when trained on a huge and diverse dataset of web pages. It is considered a general system replacing the learning objective *P(output|input)* in the original GPT with *P(output|input, task-i)*, where the model produces the different output for the same input, conditioned on a specific task—that is, GPT-2 learns multiple tasks by training the same unsupervised model. One single pre-trained model learns different abilities just through the learning objective. We see similar formulations in multi-task and meta-task settings in other studies as well. Such a shift to **Multi-Task Learning** (**MTL**) makes it possible to perform many different tasks for the same input. But how do the models determine which task to perform? They do this through zero-shot task transfer.

Compared to the original GPT, GPT-2 has no task-specific fine-tuning and is able to work in a zero-shot-task-transfer setting, where all the downstream tasks are part of predicting conditional probabilities. The task is somehow formulated within the input, and the model is expected to understand the nature of downstream tasks and provide answers accordingly. For example, for an English-to-Turkish MT task, it is conditioned not only on the input but also on the task. The input is arranged so that an English sentence is followed by a Turkish sentence, with a delimiter from which the model understands that the task is an English-to-Turkish translation.

The OpenAI team trained the GPT-3 model (see the paper *Language models are few-shot learners* (2020)) with 175 billion parameters, which is 100 times bigger than GPT-2. The architecture of GPT-2 and GPT-3 is similar, with the main differences usually being in the model size and the dataset quantity/quality. Due to the massive amount of data in the dataset and the large number of parameters it is trained on, it achieved better results on many downstream tasks in zero-shot, one-shot, and few-shot (*K=32*) settings without any gradient-based fine-tuning. The team showed that the model performance increased as the parameter size and the number of examples increased for many tasks, including translation, **Question Answering (QA)**, and masked-token tasks.

Transformer-XL

Transformer models suffer from the fixed-length context due to a lack of recurrence in the initial design and context fragmentation, although they are capable of learning long-term dependency. Most of the transformers break the documents into a list of fixed-length (mostly 512) segments, where any information flow across segments is not possible. Consequently, the language models are not able to capture long-term dependencies beyond this fixed-length limit. Moreover, the segmentation procedure builds the segments without paying attention to sentence boundaries. A segment can be absurdly made up of the second half of a sentence and the first half of its successor, hence the language models can miss the necessary contextual information when predicting the next token. This problem is referred to as *context fragmentation* problem by the studies.

To address and overcome these issues, the Transformer-XL authors (see the paper *Transformer-XL: Attentive Language Models Beyond a Fixed-Length Context* (2019)) proposed a new transformer architecture, including a segment-level recurrence mechanism and a new positional encoding scheme. This approach inspired many subsequent models. It is not limited to two consecutive segments since the effective context can extend beyond the two segments. The recurrence mechanism works between every two consecutive segments, leading to spanning the several segments to a certain degree. The largest possible dependency length that the model can attend is limited by the number of layers and segment lengths.

XLNet

Masked Language Modeling (**MLM**) dominated the pre-training phase of transformer-based architectures. However, it has faced criticism in the past since the masked tokens are present in the pre-training phase but are absent during the fine-tuning phase, which leads to a discrepancy between pre-training and fine-tuning. Because of this absence, the model may not be able to use all of the information learned during the pre-training phase. XLNet (see the paper *XLNet: Generalized Autoregressive Pretraining for Language Understanding* (2019)) replaces MLM with **Permuted Language Modeling** (**PLM**), which is a random permutation of the input tokens to overcome this bottleneck. The permutation language modeling makes each token position utilize contextual information from all positions, leading to capturing bidirectional context. The objective function only permutes the factorization order and defines the order of token predictions, but doesn't change the natural positions of sequences. Briefly, the model chooses some tokens as a target after permutation, and it further tries to predict them conditioned on the remaining tokens and the natural positions of the target. It makes it possible to use an AR model in a bidirectional fashion.

XLNet takes advantage of both AE and AR models. It is, indeed, a generalized AR model; however, it can attend the tokens from both left and right contexts, thanks to permutation-based language modeling. Besides its objective function, XLNet is made up of two important mechanisms: it integrates the segment-level recurrence mechanism of Transformer-XL into its framework, and it includes the careful design of the two-stream attention mechanism for target-aware representations.

Let's discuss the models, using both parts of the Transformers in the next section.

Working with Seq2Seq models

The left encoder and the right decoder part of the transformer are connected with cross-attention, which helps each decoder layer attend over the final encoder layer. This naturally pushes models toward producing output that closely ties to the original input. A Seq2Seq model, which is the original transformer, achieves this by using the following scheme:

Input tokens-> embeddings-> encoder-> decoder-> output tokens

Seq2Seq models keep the encoder and decoder part of the transformer. T5, **Bidirectional and Auto-Regressive Transformer** (**BART**), and **Pre-training with Extracted Gap-sentences for Abstractive Summarization Sequence-to-Sequence models** (**PEGASUS**) are among the popular Seq2Seq models.

T5

Most NLP architectures, ranging from Word2Vec to transformers learn embeddings and other parameters by predicting the masked words using context (neighbor) words. We treat NLP problems as word prediction problems. Some studies cast almost all NLP problems as QA or token classification. Likewise, T5 (see the paper *Exploring the Limits of Transfer Learning with a Unified Text-to-Text Transformer* (2019)) proposed a unifying framework to solve many tasks by casting them to a text-to-text problem. The idea underlying T5 is to cast all NLP tasks to a text-to-text (Seq2Seq) problem where both input and output are a list of tokens because the text-to-text framework has been found to be beneficial in applying the same model to diverse NLP tasks from QA to text summarization.

The following diagram, which is inspired from the original paper, shows how T5 solves four different NLP problems—MT, linguistic acceptability, semantic similarity, and summarization—within a unified framework:

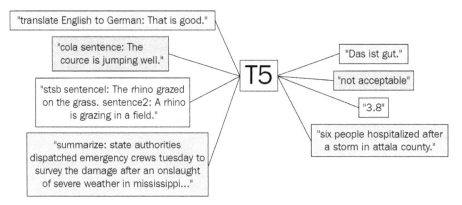

Figure 4.3 – Diagram of the T5 framework

The T5 model roughly follows the original encoder-decoder transformer model. The modifications are done in the layer normalization and position embeddings scheme. Instead of using sinusoidal positional embedding or learned embedding, T5 uses relative positional embedding, which is becoming more common in transformer architectures. T5 is a single model that can work on a diverse set of tasks such as language generation. More importantly, it casts tasks into a text-to-text format. The model is fed with text that is made up of a task prefix and the input attached to it. We convert a labeled textual dataset to a `{'inputs': '....', 'targets': ...'}` format, where we insert the purpose in the input as a prefix. Then, we train the model with labeled data so that it learns what to do and how to do it. As shown in the preceding diagram, for the English-German translation task, the `"translate English to German: That is good."` input is going to produce `"das is gut."`. Likewise, any input with a `"summarize:"` prefix will be summarized by the model.

Introducing BART

As with XLNet, the BART model (see the paper *BART: Denoising Sequence-to-Sequence Pre-training for Natural Language Generation, Translation, and Comprehension* (2019)) takes advantage of the schemes of AE and AR models. It uses standard Seq2Seq transformer architecture, with a small modification. BART is a pre-trained model using a variety of noising approaches that corrupt documents. The major contribution of the study to the field is that it allows us to apply several types of creative corruption schemes, as shown in the following diagram:

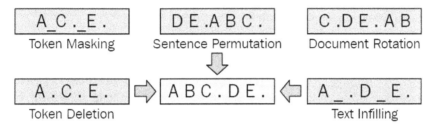

Figure 4.4 – Diagram inspired by the original BART paper

We will look at each scheme in detail, as follows:

- **Token Masking**: Tokens are randomly masked with a [MASK] symbol, the same as with the BERT model.

- **Token Deletion**: Tokens are randomly removed from the documents. The model is forced to determine which positions are removed.

- **Text Infilling**: Following SpanBERT, a number of text spans are sampled, and then they are replaced by a single [MASK] token. There is also [MASK] token insertion.

- **Sentence Permutation**: The sentences in the input are segmented and shuffled in random order.

- **Document Rotation**: The document is rotated so that it begins with a randomly selected token, **C** in the case in the preceding diagram. The objective is to find the start position of a document.

The BART model can be fine-tuned in several ways for downstream applications such as BERT. For the task of sequence classification, the input is passed through encoder and decoder, and the final hidden state of the decoder is considered the learned representation. Then, a simple linear classifier can make predictions. Likewise, for token classification tasks, the entire document is fed into the encoder and decoder, and the last state of the final decoder is the representation for each token. Based on these representations, we can solve the token classification problem, which we will discuss in *Chapter 6, Fine-Tuning Language Models for Token Classification*. **Named-Entity Recognition** (**NER**) and **Part-Of-Speech** (**POS**) tasks can be solved using this final representation, where NER identifies entities such as person and organization in a text and POS associates each token with their lexical categories, such as noun, adjective, and so on.

For sequence generation, the decoder block of the BART model, which is an AR decoder, can be directly fine-tuned for sequence-generation tasks such as abstractive QA or summarization. The BART authors (Lewis, Mike, et al.) trained the models using two standard summarization datasets: CNN/DailyMail and XSum. The authors also showed that it is possible to use both the encoder part—which consumes a source language—and the decoder part, which produces the words in the target language as a single pre-trained decoder for MT. They replaced the encoder embedding layer with a new randomly initialized encoder in order to learn words in the source language. Then, the model is trained in an end-to-end fashion, which trains the new encoder to map foreign words into an input that BART can denoise to the target language. The new encoder can use a separate vocabulary, including foreign language, from the original BART model.

In the HuggingFace platform, we can access the original pre-trained BART model with the following line of code:

```
AutoModel.from_pretrained('facebook/bart-large')
```

When we call the standard `summarization` pipeline of the `transformers` library, as shown in the following line of code, a distilled pre-trained BART model is loaded. This call implicitly loads the `"sshleifer/distilbart-cnn-12-6"` model and the corresponding tokenizers, as follows:

```
summarizer = pipeline("summarization")
```

The following code explicitly loads the same model and the corresponding tokenizer. The code example takes a text to be summarized and outputs the results:

```
from transformers import BartTokenizer,
BartForConditionalGeneration, BartConfig
from transformers import pipeline
model = \
```

```
BartForConditionalGeneration.from_pretrained('sshleifer/
distilbart-cnn-12-6')
tokenizer = BartTokenizer.from_pretrained('sshleifer/
distilbart-cnn-12-6')
nlp=pipeline("summarization", model=model, tokenizer=tokenizer)
text='''
We order two different types of jewelry from this
company the other jewelry we order is perfect.
However with this jewelry I have a few things I
don't like. The little Stone comes out of these
and customers are complaining and bringing them
back and we are having to put new jewelry in their
holes. You cannot sterilize these in an autoclave
as well because it heats up too much and the glue
does not hold up so the second group of these that
we used I did not sterilize them that way and the
stones still came out. When I use a dermal clamp
to put the top on the stones come out immediately.
DO not waste your money on this particular product
buy the three mm. that has the claws that hold the
jewelry in those are perfect. So now I'm stuck
with jewelry that I can't sell not good for
business.'''
q=nlp(text)
import pprint
pp = pprint.PrettyPrinter(indent=0, width=100)
pp.pprint(q[0]['summary_text'])
(' The little Stone comes out of these little stones and
customers are complaining and bringing ' 'them back and we are
having to put new jewelry in their holes . You cannot sterilize
these in an ' 'autoclave because it heats up too much and the
glue does not hold up so the second group of ' 'these that we
used I did not sterilize them that way and the stones still
came out .')
```

In the next section, we get our hands dirty and learn how to train such models.

AR language model training

In this section, you will learn how it is possible to train your own AR language models. We will start with GPT-2 and get a deeper look inside its different functions for training, using the `transformers` library.

You can find any specific corpus to train your own GPT-2, but for this example, we used *Emma* by Jane Austen, which is a romantic novel. Training on a much bigger corpus is highly recommended to have a more general language generation.

Before we start, it's good to note that we used TensorFlow's native training functionality to show that all Hugging Face models can be directly trained on TensorFlow or PyTorch if you wish to. Follow these steps:

1. You can download the *Emma* novel raw text by using the following command:

    ```
    wget https://raw.githubusercontent.com/teropa/nlp/master/
    resources/corpora/gutenberg/austen-emma.txt
    ```

2. The first step is to train the `BytePairEncoding` tokenizer for GPT-2 on a corpus that you intend to train your GPT-2 on. The following code will import the `BPE` tokenizer from the `tokenizers` library:

    ```
    from tokenizers.models import BPE
    from tokenizers import Tokenizer
    from tokenizers.decoders import ByteLevel as
    ByteLevelDecoder
    from tokenizers.normalizers import Sequence, Lowercase
    from tokenizers.pre_tokenizers import ByteLevel
    from tokenizers.trainers import BpeTrainer
    ```

3. As you see, in this example, we intend to train a more advanced tokenizer by adding more functionality, such as the `Lowercase` normalization. To make a `tokenizer` object, you can use the following code:

    ```
    tokenizer = Tokenizer(BPE())
    tokenizer.normalizer = Sequence([
        Lowercase()
    ])
    tokenizer.pre_tokenizer = ByteLevel()
    tokenizer.decoder = ByteLevelDecoder()
    ```

The first line makes a tokenizer from the BPE tokenizer class. For the normalization part, Lowercase has been added, and the pre_tokenizer attribute is set to be as ByteLevel to ensure we have bytes as our input. The decoder attribute must be also set to ByteLevelDecoder to be able to decode correctly.

4. Next, the tokenizer will be trained using a 50000 maximum vocabulary size and an initial alphabet from ByteLevel, as follows:

```
trainer = BpeTrainer(vocab_size=50000, inital_
alphabet=ByteLevel.alphabet(), special_tokens=[
            "<s>",
            "<pad>",
            "</s>",
            "<unk>",
            "<mask>"
      ])
tokenizer.train(["austen-emma.txt"], trainer)
```

5. It is also necessary to add special tokens to be considered. To save the tokenizer, you are required to create a directory, as follows:

```
!mkdir tokenizer_gpt
```

6. You can save the tokenizer by running the following command:

```
tokenizer.save("tokenizer_gpt/tokenizer.json")
```

7. Now that the tokenizer is saved, it's time to preprocess the corpus and make it ready for GPT-2 training using the saved tokenizer, but first, important imports must not be forgotten. The code to do the imports is illustrated in the following snippet:

```
from transformers import GPT2TokenizerFast, GPT2Config,
TFGPT2LMHeadModel
```

8. And the tokenizer can be loaded by using GPT2TokenizerFast, as follows:

```
tokenizer_gpt = GPT2TokenizerFast.from_
pretrained("tokenizer_gpt")
```

9. It is also essential to add special tokens with their marks, like this:

```
tokenizer_gpt.add_special_tokens({
    "eos_token": "</s>",
    "bos_token": "<s>",
```

```
    "unk_token": "<unk>",
    "pad_token": "<pad>",
    "mask_token": "<mask>"
})
```

10. You can also double-check to see if everything is correct or not by running the following code:

```
tokenizer_gpt.eos_token_id
>> 2
```

This code will output the **End-of-Sentence (EOS)** token **Identifier (ID)**, which is 2 for the current tokenizer.

11. You can also test it for a sentence by executing the following code:

```
tokenizer_gpt.encode("<s> this is </s>")
>> [0, 265, 157, 56, 2]
```

For this output, 0 is the beginning of the sentence, 265, 157, and 56 are related to the sentence itself, and the EOS is marked as 2, which is </s>.

12. These settings must be used when creating a configuration object. The following code will create a config object and the TensorFlow version of the GPT-2 model:

```
config = GPT2Config(
    vocab_size=tokenizer_gpt.vocab_size,
    bos_token_id=tokenizer_gpt.bos_token_id,
    eos_token_id=tokenizer_gpt.eos_token_id
)
model = TFGPT2LMHeadModel(config)
```

13. On running the config object, you can see the configuration in dictionary format, as follows:

```
config
>> GPT2Config {  "activation_function": "gelu_new",
"attn_pdrop": 0.1,   "bos_token_id": 0,   "embd_pdrop":
0.1,   "eos_token_id": 2,   "gradient_checkpointing":
false,   "initializer_range": 0.02,   "layer_norm_
epsilon": 1e-05,   "model_type": "gpt2",   "n_ctx": 1024,
"n_embd": 768,   "n_head": 12,   "n_inner": null,   "n_
layer": 12,   "n_positions": 1024,   "resid_pdrop": 0.1,
"summary_activation": null,   "summary_first_dropout":
```

```
0.1,  "summary_proj_to_labels": true,  "summary_type":
"cls_index",  "summary_use_proj": true,  "transformers_
version": "4.3.2",  "use_cache": true,  "vocab_size":
11750}
```

As you can see, other settings are not touched, and the interesting part is that vocab_size is set to 11750. The reason behind this is that we set the maximum vocabulary size to be 50000, but the corpus had less, and its **Byte-Pair Encoding (BPE)** token created 11750.

14. Now, you can get your corpus ready for pre-training, as follows:

```
with open("austen-emma.txt", "r", encoding='utf-8') as f:
    content = f.readlines()
```

15. The content will now include all raw text from the raw file, but it is required to remove '\n' from each line and drop lines with fewer than 10 characters, as follows:

```
content_p = []
for c in content:
    if len(c)>10:
        content_p.append(c.strip())
content_p = " ".join(content_p)+tokenizer_gpt.eos_token
```

16. Dropping short lines will ensure that the model is trained on long sequences, to be able to generate longer sequences. At the end of the preceding snippet, content_p has the concatenated raw file with eos_token added to the end. But you can follow different strategies too—for example, you can separate each line by adding </s> to each line, which will help the model to recognize when the sentence ends. However, we intend to make it work for much longer sequences without encountering EOS. The code is illustrated in the following snippet:

```
tokenized_content = tokenizer_gpt.encode(content_p)
```

The GPT tokenizer from the preceding code snippet will tokenize the whole text and make it one whole, long sequence of token IDs.

17. Now, it's time to make the samples for training, as follows:

```
sample_len = 100
examples = []
for i in range(0, len(tokenized_content)):
    examples.append(tokenized_content[i:i + sample_len])
```

18. The preceding code makes `examples` a size of `100` for each one starting from a given part of text and ending at `100` tokens later:

```
train_data = []
labels = []
for example in examples:
    train_data.append(example[:-1])
    labels.append(example[1:])
```

In `train_data`, there will be a sequence of size `99` from start to the 99th token, and the labels will have a token sequence from `1` to `100`.

19. For faster training, it is required to make the data in the form of a TensorFlow dataset, as follows:

```
Import tensorflow as tf
buffer = 500
batch_size = 16
dataset = tf.data.Dataset.from_tensor_slices((train_data,
labels))
dataset = dataset.shuffle(buffer).batch(batch_size, drop_
remainder=True)
```

`buffer` is the buffer size used for shuffling data, and `batch_size` is the batch size for training. `drop_remainder` is used to drop the remainder if it is less than `16`.

20. Now, you can specify your `optimizer`, `loss`, and `metrics` properties, as follows:

```
optimizer = tf.keras.optimizers.Adam(learning_rate=3e-5,
epsilon=1e-08, clipnorm=1.0)
loss = tf.keras.losses.SparseCategoricalCrossentropy(from_
logits=True)
metric = tf.keras.metrics.
SparseCategoricalAccuracy('accuracy')
model.compile(optimizer=optimizer, loss=[loss, *[None] *
model.config.n_layer], metrics=[metric])
```

21. And the model is compiled and ready to be trained with the number of epochs you wish, as follows:

```
epochs = 10
model.fit(dataset, epochs=epochs)
```

You will see an output that looks something like this:

```
Epoch 1/10
WARNING:tensorflow:The parameters `output_attentions`, `output_hidden_states` and `use_cache` cannot be updated wh
en calling a model.They have to be set to True/False in the config object (i.e.: `config=XConfig.from_pretrained('
name', output_attentions=True)`).
WARNING:tensorflow:The parameter `return_dict` cannot be set in graph mode and will always be set to `True`.
WARNING:tensorflow:The parameters `output_attentions`, `output_hidden_states` and `use_cache` cannot be updated wh
en calling a model.They have to be set to True/False in the config object (i.e.: `config=XConfig.from_pretrained('
name', output_attentions=True)`).
WARNING:tensorflow:The parameter `return_dict` cannot be set in graph mode and will always be set to `True`.
166/166 [==============================] - 421s 3s/step - loss: 5.8450 - logits_loss: 5.8450 - logits_accuracy: 0.
1649 - past_key_values_1_accuracy: 0.0025 - past_key_values_2_accuracy: 0.0021 - past_key_values_3_accuracy: 0.001
7 - past_key_values_4_accuracy: 0.0031 - past_key_values_5_accuracy: 0.0024 - past_key_values_6_accuracy: 0.0026 -
past_key_values_7_accuracy: 0.0029 - past_key_values_8_accuracy: 0.0030 - past_key_values_9_accuracy: 0.0032 - pas
t_key_values_10_accuracy: 0.0028 - past_key_values_11_accuracy: 0.0015 - past_key_values_12_accuracy: 0.0039
Epoch 2/10
166/166 [==============================] - 421s 3s/step - loss: 2.6242 - logits_loss: 2.6242 - logits_accuracy: 0.
5361 - past_key_values_1_accuracy: 0.0025 - past_key_values_2_accuracy: 0.0023 - past_key_values_3_accuracy: 0.002
3 - past_key_values_4_accuracy: 0.0025 - past_key_values_5_accuracy: 0.0025 - past_key_values_6_accuracy: 0.0025 -
past_key_values_7_accuracy: 0.0024 - past_key_values_8_accuracy: 0.0024 - past_key_values_9_accuracy: 0.0027 - pas
t_key_values_10_accuracy: 0.0027 - past_key_values_11_accuracy: 0.0025 - past_key_values_12_accuracy: 0.0025
Epoch 3/10
 42/166 [======>.......................] - ETA: 5:07 - loss: 1.4982 - logits_loss: 1.4982 - logits_accuracy: 0.760
7 - past_key_values_1_accuracy: 0.0027 - past_key_values_2_accuracy: 0.0023 - past_key_values_3_accuracy: 0.0022 -
past_key_values_4_accuracy: 0.0026 - past_key_values_5_accuracy: 0.0025 - past_key_values_6_accuracy: 0.0023 - pas
t_key_values_7_accuracy: 0.0024 - past_key_values_8_accuracy: 0.0023 - past_key_values_9_accuracy: 0.0027 - past_k
ey_values_10_accuracy: 0.0027 - past_key_values_11_accuracy: 0.0025 - past_key_values_12_accuracy: 0.0023
```

Figure 4.5 – GPT-2 training using TensorFlow/Keras

We will now look at NLG using AR models. Now that you have saved the model, it will be used for generating sentences in the next section.

Up until this point, you have learned how it is possible to train your own model for NLG. In the next section, we describe how to utilize NLG models for language generation.

NLG using AR models

In the previous section, you have learned how it is possible to train an AR model on your own corpus. As a result, you have trained the GPT-2 version of your own. But the missing answer to the question *How can I use it?* remains. To answer that, let's proceed as follows:

1. Let's start generating sentences from the model you have just trained, as follows:

```
def generate(start, model):
    input_token_ids = tokenizer_gpt.encode(start, return_
tensors='tf')
    output = model.generate(
        input_token_ids,
```

```
        max_length = 500,
        num_beams = 5,
        temperature = 0.7,
        no_repeat_ngram_size=2,
        num_return_sequences=1
    )
    return tokenizer_gpt.decode(output[0])
```

The `generate` function that is defined in the preceding code snippet takes a `start` string and generates sequences following that string. You can change parameters such as `max_length` to be set to a smaller sequence size or `num_return_sequences` to have different generations.

2. Let's just try it with an empty string, as follows:

```
generate(" ", model)
```

We get the following output:

```
' it was a nervous; and emma could not but it, and made it necessary to be cheerful. his spirits required with th
em; hating change of every kind.  he was by no means yet reconciled to his own daughter\'s marrying, was always di
sagreeable; but with compassion, as the origin of affection, though it had been entirely a mile from his habits of
being never able to part with miss taylor been a great must be accepted in herself as for herself; fond of gentle
selfishness, nor could ever speak of her own, only half a long october and from isabella\'s being now obliged to s
uppose that great deal happier if she was very much beyond her father and he could feel differently from such a go
od was now long ago for him from himself to have had done as sad a thing for her life at hartfield. emma was much
older man in their little too long evening, when and would not have been living together as her friend and a very
early from them, they had spent all the house of having been supplied by any means rank as cheerfully as friend wa
s her but emma smiled and bear the rest of emma had ceased to say her sister\'s marriage, "how she had died too mu
ch disposed to keep him not to think a little children of authority being settled in london was some satisfaction
in great danger of great comfort and chatted as large and with what a house from five years had said at any time.
weston was more than an excellent woman as he had many a man, the want of the with her temper had her advantages,
but little way and her daily but particularly was no companion for even half so unperceived, who had taught and wi
sh she dearly loved her through the actual disparity in mr. woodhouse had such an affection; you have never any re
straint as governess than any odd humours, it is such thoughts; highly her!" "i cannot that other-and you know fro
m a large.--and how was used to see her in affection for having great house; these were first with all his heart a
nd had hardly allowed her pleasant society again. how nursed her many now in consequence of a much have recommende
d him at this is the friendliness of sixteen years old and friend very mutually attached, being left to impose any
disagreeable consciousness were here to sit and november evening must go in the next visit from intellectual such
as few a match of both daughters, very good-people have more the advantage of his life in ways than a melancholy c
hange than her as great with you would be felt every body that her place had miss'
```

Figure 4.6 – GPT-2 text-generation example

As you can see from the preceding output, a long text is generated, even if the semantics of the text is not very pleasing, but the syntax is almost correct in many cases.

3. Now, let's try different starts, with `max_length` set to a lower value such as 3 0, as follows:

```
generate("wetson was very good")
>> 'wetson was very good; but it, that he was a great
must be a mile from them, and a miss taylor in the
house;'
```

As you recall `weston` is one of the characters from the novel.

4. To save the model, you can use the following code to make it reusable for publishing or different applications:

```
model.save_pretrained("my_gpt-2/")
```

5. To make sure your model is saved correctly, you can try loading it, as follows:

```
model_reloaded = TFGPT2LMHeadModel.from_pretrained("my_
gpt-2/")
```

Two files are saved—a `config` file and a `model.h5` file, which is for the TensorFlow version. We can see both of these files in the following screenshot:

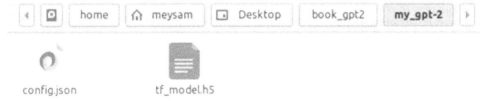

Figure 4.7 – Language model save_pretrained output

6. Hugging Face also has a standard for filenames that must be used—these standard filenames are available by using the following import:

```
from transformers import WEIGHTS_NAME, CONFIG_NAME, TF2_
WEIGHTS_NAME
```

However, when using the `save_pretrained` function, it is not required to put the filenames—just the directory will suffice.

7. Hugging Face also has `AutoModel` and `AutoTokenizer` classes, as you have seen from the previous sections. You can also use this functionality to save the model, but before doing that there are still a few configurations that need to be done manually. The first thing is to save the tokenizer in the proper format to be used by `AutoTokenizer`. You can do this by using `save_pretrained`, as follows:

```
tokenizer_gpt.save_pretrained("tokenizer_gpt_auto/")
```

This is the output:

```
('tokenizer_gpt_auto/tokenizer_config.json',
 'tokenizer_gpt_auto/special_tokens_map.json',
 'tokenizer_gpt_auto/vocab.json',
 'tokenizer_gpt_auto/merges.txt',
 'tokenizer_gpt_auto/added_tokens.json')
```

Figure 4.8 – Tokenizer save_pretrained output

8. The file list is shown in the directory you specified, but `tokenizer_config` must be manually changed to be usable. First, you should rename it as `config.json`, and secondly, you should add a property in **JavaScript Object Notation (JSON)** format, indicating that the `model_type` property is `gpt2`, as follows:

    ```
    {"model_type":"gpt2",
    ...
    }
    ```

9. Now, everything is ready, and you can simply use these two lines of code to load `model` and `tokenizer`:

    ```
    model = AutoModel.from_pretrained("my_gpt-2/", from_tf=True)
    tokenizer = AutoTokenizer.from_pretrained("tokenizer_gpt_auto")
    ```

    However, do not forget to set `from_tf` to `True` because your model is saved in TensorFlow format.

Up to this point, you have learned how you can pre-train and save your own text-generation model using `tensorflow` and `transformers`. You also learned how it is possible to save a pre-trained model and prepare it to be used as an auto model. In the next section, you will learn the basics of using other models.

Summarization and MT fine-tuning using simpletransformers

Up to now, you have learned the basics and advanced methods of training language models, but it is not always feasible to train your own language model from scratch because there are sometimes impediments such as low computational power. In this section, you will look at how to fine-tune language models on your own datasets for specific tasks of MT and summarization. Follow these next steps:

1. To start, you need to install the `simpletransformers` library, as follows:

    ```
    pip install simpletransformers
    ```

2. The next step is to download the dataset that contains your parallel corpus. This parallel corpus can be of any type of Seq2Seq task. For this example, we are going to use the MT example, but you can use any other dataset for other tasks such as paraphrasing, summarization, or even for converting text to **Structured Query Language** (**SQL**).

    You can download the dataset from `https://www.kaggle.com/seymasa/turkish-to-english-translation-dataset/version/1`.

3. After you have downloaded and unpacked the data, it is necessary to add EN and TR for column headers, for easier use. You can load the dataset using `pandas`, as follows:

    ```
    import pandas as pd
    df = pd.read_csv("TR2EN.txt",sep="\t").astype(str)
    ```

4. It is required to add T5-specific commands to the dataset to make it understand the command it is dealing with. You can do this with the following code:

    ```
    data = []
    for item in digitrons():
        data.append(["translate english to turkish", item[1].
    EN, item[1].TR])
    ```

5. Afterward, you can reform the DataFrame, like this:

    ```
    df = pd.DataFrame(data, columns=["prefix", "input_text",
    "target_text"])
    ```

The result is shown in the following screenshot:

	prefix	input_text	target_text
0	translate english to turkish	Hi.	Merhaba.
1	translate english to turkish	Hi.	Selam.
2	translate english to turkish	Run!	Kaç!
3	translate english to turkish	Run!	Koş!
4	translate english to turkish	Run.	Kaç!
5	translate english to turkish	Run.	Koş!
6	translate english to turkish	Who?	Kim?
7	translate english to turkish	Fire!	Ateş!
8	translate english to turkish	Fire!	Yangın!
9	translate english to turkish	Help!	Yardım et!
10	translate english to turkish	Jump.	Defol.
11	translate english to turkish	Stop!	Dur!
12	translate english to turkish	Stop!	Bırak!
13	translate english to turkish	Wait.	Bekle.

Figure 4.9 – English-Turkish MT parallel corpus

6. Next, run the following code to import the required classes:

```
from simpletransformers.t5 import T5Model, T5Args
```

7. Defining arguments for training is accomplished using the following code:

```
model_args = T5Args()
model_args.max_seq_length = 96
model_args.train_batch_size = 20
model_args.eval_batch_size = 20
model_args.num_train_epochs = 1
model_args.evaluate_during_training = True
```

```
model_args.evaluate_during_training_steps = 30000
model_args.use_multiprocessing = False
model_args.fp16 = False
model_args.save_steps = -1
model_args.save_eval_checkpoints = False
model_args.no_cache = True
model_args.reprocess_input_data = True
model_args.overwrite_output_dir = True
model_args.preprocess_inputs = False
model_args.num_return_sequences = 1
model_args.wandb_project = "MT5 English-Turkish
Translation"
```

8. At the end, you can load any model you wish to fine-tune. Here's the one we've chosen:

    ```
    model = T5Model("mt5", "google/mt5-small", args=model_
    args, use_cuda=False)
    ```

    Don't forget to set use_cuda to False if you do not have enough **Compute Unified Device Architecture (CUDA)** memory for mT5.

9. Splitting the train and eval DataFrames can be done using the following code:

    ```
    train_df = df[: 470000]
    eval_df = df[470000:]
    ```

10. The last step is to use the following code to start training:

    ```
    model.train_model(train_df, eval_data=eval_df)
    ```

    The result of the training will be shown, as follows:

    ```
    (3,
     {'global_step': [3],
      'eval_loss': [28.536166508992512],
      'train_loss': [33.57326889038086]})
    ```

    Figure 4.10 – mT5 model evaluation results

    This indicates evaluation and training loss.

11. You can simply load and use the model with the following code:

    ```
    model_args = T5Args()
    model_args.max_length = 512
    ```

```
model_args.length_penalty = 1
model_args.num_beams = 10
model = T5Model("mt5", "outputs", args=model_args, use_
cuda=False)
```

The `model_predict` function can be used now for the translation from English to Turkish.

The Simple Transformers library (`simpletransformers`) makes training many models, from sequence labeling to Seq2Seq models, very easy and usable.

Well done! We have learned how to train our own AR models and have come to the end of this chapter.

Summary

In this chapter, we have learned various aspects of AR language models, from pre-training to fine-tuning. We looked at the best features of such models by training generative language models and fine-tuning on tasks such as MT. We understood the basics of more complex models such as T5 and used this kind of model to perform MT. We also used the `simpletransformers` library. We trained GPT-2 on our own corpus and generated text using it. We learned how to save it and use it with `AutoModel`. We also had a deeper look into how BPE can be trained and used, using the `tokenizers` library.

In the next chapter, we will see how to fine-tune models for text classification.

References

Here are a few references that you can use to expand on what we learned in this chapter:

- *Radford, A., Wu, J., Child, R., Luan, D., Amodei, D.* and *Sutskever, I. (2019). Language Models are Unsupervised Multitask Learners. OpenAI blog, 1(8), 9.*

- *Lewis, M., Liu, Y., Goyal, N., Ghazvininejad, M., Mohamed, A., Levy, O.* and *Zettlemoyer, L. (2019). BART: Denoising Sequence-to-Sequence Pre-training for Natural Language Generation, Translation, and Comprehension. arXiv preprint arXiv:1910.13461.*

- *Xue, L., Constant, N., Roberts, A., Kale, M., Al-Rfou, R., Siddhant, A.* and *Raffel, C. (2020). mT5: A massively multilingual pre-trained text-to-text transformer. arXiv preprint arXiv:2010.11934.*

- *Raffel, C. , Shazeer, N. , Roberts, A. , Lee, K. , Narang, S. , Matena, M. and Liu, P. J. (2019). Exploring the Limits of Transfer Learning with a Unified Text-to-Text Transformer. arXiv preprint arXiv:1910.10683.*

- *Yang, Z., Dai, Z., Yang, Y., Carbonell, J., Salakhutdinov, R. and Le, Q. V. (2019). XLNet: Generalized Autoregressive Pretraining for Language Understanding. arXiv preprint arXiv:1906.08237.*

- *Dai, Z., Yang, Z., Yang, Y., Carbonell, J., Le, Q. V. and Salakhutdinov, R. (2019). Transformer-xl: Attentive Language Models Beyond a Fixed-Length Context. arXiv preprint arXiv:1901.02860.*

5
Fine-Tuning Language Models for Text Classification

In this chapter, we will learn how to configure a pre-trained model for text classification and how to fine-tune it to any text classification downstream task, such as sentiment analysis or multi-class classification. We will also discuss how to handle sentence-pair and regression problems by covering an implementation. We will work with well-known datasets such as GLUE, as well as our own custom datasets. We will then take advantage of the Trainer class, which deals with the complexity of processes for training and fine-tuning.

First, we will learn how to fine-tune single-sentence binary sentiment classification with the Trainer class. Then, we will train for sentiment classification with native PyTorch without the Trainer class. In multi-class classification, more than two classes will be taken into consideration. We will have seven class classification fine-tuning tasks to perform. Finally, we will train a text regression model to predict numerical values with sentence pairs.

The following topics will be covered in this chapter:

- Introduction to text classification
- Fine-tuning the BERT model for single-sentence binary classification
- Training a classification model with native PyTorch
- Fine-tuning BERT for multi-class classification with custom datasets
- Fine-tuning BERT for sentence-pair regression
- Utilizing `run_glue.py` to fine-tune the models

Technical requirements

We will be using Jupyter Notebook to run our coding exercises. You will need Python 3.6+ for this. Ensure that the following packages are installed:

- `sklearn`
- Transformers 4.0+
- `datasets`

All the notebooks for the coding exercises in this chapter will be available at the following GitHub link: `https://github.com/PacktPublishing/Mastering-Transformers/tree/main/CH05`.

Check out the following link to see the Code in Action video:

`https://bit.ly/3y5Fe6R`

Introduction to text classification

Text classification (also known as text categorization) is a way of mapping a document (sentence, Twitter post, book chapter, email content, and so on) to a category out of a predefined list (classes). In the case of two classes that have positive and negative labels, we call this **binary classification** – more specifically, **sentiment analysis**. For more than two classes, we call this **multi-class classification**, where the classes are mutually exclusive, or **multi-label classification**, where the classes are not mutually exclusive, which means a document can receive more than one label. For instance, the content of a news article may be related to sport and politics at the same time. Beyond this classification, we may want to score the documents in a range of [-1,1] or rank them in a range of [1-5]. We can solve this kind of problem with a regression model, where the type of the output is numeric, not categorical.

Luckily, the transformer architecture allows us to efficiently solve these problems. For sentence-pair tasks such as document similarity or textual entailment, the input is not a single sentence, but rather two sentences, as illustrated in the following diagram. We can score to what degree two sentences are semantically similar or predict whether they are semantically similar. Another sentence-pair task is **textual entailment**, where the problem is defined as multi-class classification. Here, two sequences are consumed in the GLUE benchmark: entail/contradict/neutral:

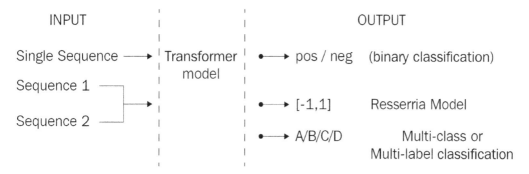

Figure 5.1 – Text classification scheme

Let's start our training process by fine-tuning a pre-trained BERT model for a common problem: sentiment analysis.

Fine-tuning a BERT model for single-sentence binary classification

In this section, we will discuss how to fine-tune a pre-trained BERT model for sentiment analysis by using the popular `IMDb sentiment` dataset. Working with a GPU will speed up our learning process, but if you do not have such resources, you can work with a CPU as well for fine-tuning. Let's get started:

1. To learn about and save our current device, we can execute the following lines of code:

    ```
    from torch import cuda
    device = 'cuda' if cuda.is_available() else 'cpu'
    ```

2. We will use the `DistilBertForSequenceClassification` class here, which is inherited from the `DistilBert` class, with a special sequence classification head at the top. We can utilize this *classification head* to train the classification model, where the number of classes is 2 by default:

    ```
    from transformers import DistilBertTokenizerFast,
    DistilBertForSequenceClassification
    model_path= 'distilbert-base-uncased'
    tokenizer = DistilBertTokenizerFast.from_
    pre-trained(model_path)
    model = \ DistilBertForSequenceClassification.from_
    pre-trained(model_path, id2label={0:"NEG", 1:"POS"},
    label2id={"NEG":0, "POS":1})
    ```

3. Notice that two parameters called `id2label` and `label2id` are passed to the model to use during inference. Alternatively, we can instantiate a particular `config` object and pass it to the model, as follows:

    ```
    config = AutoConfig.from_pre-trained(....)
    SequenceClassification.from_pre-trained(....
    config=config)
    ```

4. Now, let's select a popular sentiment classification dataset called `IMDB Dataset`. The original dataset consists of two sets of data: 25,000 examples for training and 25 examples for testing. We will split the dataset into test and validation sets. Note that the examples for the first half of the dataset are positive, while the second half's examples are all negative. We can distribute the examples as follows:

```
from datasets import load_dataset
imdb_train= load_dataset('imdb', split="train")
imdb_test= load_dataset('imdb', split="test[:6250]+t
est[-6250:]")
imdb_val= \
load_dataset('imdb', split="test[6250:12500]+t
est[-12500:-6250]")
```

5. Let's check the shape of the dataset:

```
>>> imdb_train.shape, imdb_test.shape, imdb_val.shape
((25000, 2), (12500, 2), (12500, 2))
```

6. You can take a small portion of the dataset based on your computational resources. For a smaller portion, you should run the following code to select 4,000 examples for training, 1,000 for testing, and 1,000 for validation, like so:

```
imdb_train= load_dataset('imdb', split="train[:2000]+tr
ain[-2000:]")
imdb_test= load_dataset('imdb',
split="test[:500]+test[-500:]")
imdb_val= load_dataset('imdb', split="test[500:1000]+t
est[-1000:-500]")
```

7. Now, we can pass these datasets through the `tokenizer` model to make them ready for training:

```
enc_train = imdb_train.map(lambda e: tokenizer(
e['text'], padding=True, truncation=True), batched=True,
batch_size=1000)
enc_test =  imdb_test.map(lambda e: tokenizer( e['text'],
padding=True, truncation=True), batched=True, batch_
size=1000)
enc_val =  imdb_val.map(lambda e: tokenizer( e['text'],
padding=True, truncation=True), batched=True, batch_
size=1000)
```

8. Let's see what the training set looks like. The attention mask and input IDs were added to the dataset by the tokenizer so that the BERT model can process:

```
import pandas as pd
pd.DataFrame(enc_train)
```

The output is as follows:

	attention_mask	input_ids	label	text
0	[1, 1, 1, 1, 1, 1, 1, 1, 1, 1, 1, 1, 1, 1, ...	[101, 22953, 2213, 4381, 2152, 2003, 1037, 947...	1	Bromwell High is a cartoon comedy. It ran at t...
1	[1, 1, 1, 1, 1, 1, 1, 1, 1, 1, 1, 1, 1, 1, ...	[101, 11573, 2791, 1006, 2030, 2160, 24913, 20...	1	Homelessness (or Houselessness as George Carli...
2	[1, 1, 1, 1, 1, 1, 1, 1, 1, 1, 1, 1, 1, 1, ...	[101, 8235, 2058, 1011, 3772, 2011, 23920, 575...	1	Brilliant over-acting by Lesley Ann Warren. Be...
3	[1, 1, 1, 1, 1, 1, 1, 1, 1, 1, 1, 1, 1, 1, ...	[101, 2023, 2003, 4089, 1996, 2087, 2104, 9250...	1	This is easily the most underrated film inn th...
4	[1, 1, 1, 1, 1, 1, 1, 1, 1, 1, 1, 1, 1, 1, ...	[101, 2023, 2003, 2025, 1996, 5171, 11463, 837...	1	This is not the typical Mel Brooks film. It wa...
...	...	...	...	...
24995	[1, 1, 1, 1, 1, 1, 1, 1, 1, 1, 1, 1, 1, 1, ...	[101, 2875, 1996, 2203, 1997, 1996, 3185, 1010...	0	Towards the end of the movie, I felt it was to...
24996	[1, 1, 1, 1, 1, 1, 1, 1, 1, 1, 1, 1, 1, 1, ...	[101, 2023, 2003, 1996, 2785, 1997, 3185, 2008...	0	This is the kind of movie that my enemies cont...
24997	[1, 1, 1, 1, 1, 1, 1, 1, 1, 1, 1, 1, 1, 1, ...	[101, 1045, 2387, 1005, 6934, 1005, 2197, 2305...	0	I saw 'Descent' last night at the Stockholm Fi...
24998	[1, 1, 1, 1, 1, 1, 1, 1, 1, 1, 1, 1, 1, 1, ...	[101, 2070, 3152, 2008, 2017, 4060, 2039, 2005...	0	Some films that you pick up for a pound turn o...
24999	[1, 1, 1, 1, 1, 1, 1, 1, 1, 1, 1, 1, 1, 1, ...	[101, 2023, 2003, 2028, 1997, 1996, 12873, 435...	0	This is one of the dumbest films, I've ever se...

25000 rows × 4 columns

Figure 5.2 – Encoded training dataset

At this point, the datasets are ready for training and testing. The `Trainer` class (`TFTrainer` for TensorFlow) and the `TrainingArguments` class (`TFTrainingArguments` for TensorFlow) will help us with much of the training complexity. We will define our argument set within the `TrainingArguments` class, which will then be passed to the `Trainer` object.

Let's define what each training argument does:

Argument	Definition
output_dir	Points to the model checkpoints and where the predictions will be saved at the end.
do_train and do_eval	Options for monitoring the model's performance during training.
logging_strategy	The options here are no, epoch, and steps (by default).
logging_steps (default value: 500)	The number of steps between two logs to be saved into the `logging_dir` director.
save_strategy	This is used to save the model checkpoint. The options are `no`, `epoch`, and `steps` (default).
save_steps (def. 500)	The number of steps between two checkpoints.
fp16	This is for mixed precision and uses both 16-bit and 32-bit floating-point types to make the model train faster and use less memory.
load_best_model_at_end	As the name suggests, this will load the best model checkpoint in terms of validation loss at the end of training.
logging_dir	TensorBoard log directory.

Table 1 – Table of different training argument definitions

9. For more information, please check the API documentation of `TrainingArguments` or execute the following code in a Python notebook:

```
TrainingArguments?
```

10. Although deep learning architectures such as LSTM need many epochs, sometimes more than 50, for transformer-based fine-tuning, we will typically be satisfied with an epoch number of 3 due to transfer learning. Most of the time, this number is enough for fine-tuning, as a pre-trained model learns a lot about the language during the pre-training phase, which takes about 50 epochs on average. To determine the correct number of epochs, we need to monitor training and evaluation loss. We will learn how to track training in *Chapter 11, Attention Visualization and Experiment Tracking*.

11. This will be enough for many downstream task problems, as we will see here. During the training process, our model checkpoints will be saved under the `./MyIMDBModel` folder for every 200 steps:

```
from transformers import TrainingArguments, Trainer
training_args = TrainingArguments(
    output_dir='./MyIMDBModel',
    do_train=True,
    do_eval=True,
    num_train_epochs=3,
    per_device_train_batch_size=32,
    per_device_eval_batch_size=64,
    warmup_steps=100,
    weight_decay=0.01,
    logging_strategy='steps',
    logging_dir='./logs',
    logging_steps=200,
    evaluation_strategy= 'steps',
        fp16= cuda.is_available(),
    load_best_model_at_end=True
)
```

12. Before instantiating a `Trainer` object, we will define the `compute_metrics()` method, which helps us monitor the progress of the training in terms of particular metrics for whatever we need, such as Precision, RMSE, Pearson correlation, BLEU, and so on. Text classification problems (such as sentiment classification or multi-class classification) are mostly evaluated with **micro-averaging** or **macro-averaging F1**. While the macro-averaging method gives equal weight to each class, micro-averaging gives equal weight to each per-text or per-token classification decision. Micro-averaging is equal to the ratio of the number of times the model decides correctly to the total number of decisions that have been made. On the other hand, the macro-averaging method computes the average score of Precision, Recall, and F1 for each class. For our classification problem, macro-averaging is more convenient for evaluation since we want to give equal weight to each label, as follows:

```
from sklearn.metrics import accuracy_score, Precision_
Recall_fscore_support
def compute_metrics(pred):
```

```
    labels = pred.label_ids
    preds = pred.predictions.argmax(-1)
    Precision, Recall, f1, _ = \
    Precision_Recall_fscore_support(labels, preds,
average='macro')
    acc = accuracy_score(labels, preds)
    return {
        'Accuracy': acc,
        'F1': f1,
        'Precision': Precision,
        'Recall': Recall
    }
```

13. We are almost ready to start the training process. Now, let's instantiate the `Trainer` object and start it. The `Trainer` class is a very powerful and optimized tool for organizing complex training and evaluation processes for PyTorch and TensorFlow (`TFTrainer` for TensorFlow) thanks to the `transformers` library:

```
trainer = Trainer(
    model=model,
    args=training_args,
    train_dataset=enc_train,
    eval_dataset=enc_val,
    compute_metrics= compute_metrics
)
```

14. Finally, we can start the training process:

```
results=trainer.train()
```

The preceding call starts logging metrics, which we will discuss in more detail in *Chapter 11, Attention Visualization and Experiment Tracking*. The entire IMDb dataset includes 25,000 training examples. With a batch size of 32, we have 25K/32 ~=782 steps, and 2,346 (782 x 3) steps to go for 3 epochs, as shown in the following progress bar:

Step	Training Loss	Validation Loss	Accuracy	F1	Precision	Recall	Runtime	Samples Per Second
		[2346/2346 21:13, Epoch 3/3]						
200	0.417800	0.239647	0.900160	0.899943	0.903660	0.900160	58.657100	213.103000
400	0.251100	0.207064	0.918960	0.918960	0.918960	0.918960	58.724400	212.859000
600	0.237300	0.188785	0.926560	0.926554	0.926707	0.926560	58.727300	212.848000
800	0.209200	0.234559	0.923680	0.923621	0.924982	0.923680	58.750400	212.764000
1000	0.128500	0.248400	0.927280	0.927280	0.927286	0.927280	58.717100	212.885000
1200	0.137400	0.251818	0.920000	0.919869	0.922771	0.920000	58.713500	212.898000
1400	0.125900	0.186671	0.930720	0.930707	0.931054	0.930720	58.724900	212.857000
1600	0.111800	0.230385	0.932960	0.932959	0.932980	0.932960	58.695400	212.964000
1800	0.051300	0.255035	0.933440	0.933440	0.933440	0.933440	58.840300	212.440000
2000	0.045200	0.269209	0.934800	0.934795	0.934927	0.934800	58.819400	212.515000
2200	0.053700	0.242861	0.934640	0.934639	0.934661	0.934640	58.836100	212.455000

The minimum loss

Figure 5.3 – The output produced by the Trainer object

15. The `Trainer` object keeps the checkpoint whose validation loss is the smallest at the end. It selects the checkpoint at step 1,400 since the validation loss at this step is the minimum. Let's evaluate the best checkpoint on three (train/test/validation) datasets:

```
>>> q=[trainer.evaluate(eval_dataset=data) for data in
[enc_train, enc_val, enc_test]]
>>> pd.DataFrame(q, index=["train","val","test"]).
iloc[:,:5]
```

The output is as follows:

	eval_loss	eval_accuracy	eval_f1	eval_precision	eval_recall
		[391/391 10:03]			
train	0.057059	0.98320	0.983199	0.983259	0.98320
val	0.186671	0.93072	0.930707	0.931054	0.93072
test	0.213239	0.92616	0.926128	0.926904	0.92616

Figure 5.4 – Classification model's performance on the train/validation/test dataset

16. Well done! We have successfully completed the training/testing phase and received 92.6 accuracy and 92.6 F1 for our macro-average. To monitor your training process in more detail, you can call advanced tools such as TensorBoard. These tools parse the logs and enable us to track various metrics for comprehensive analysis. We've already logged the performance and other metrics under the ./logs folder. Just running the tensorboard function within our Python notebook will be enough, as shown in the following code block (we will discuss TensorBoard and other monitoring tools in *Chapter 11*, *Attention Visualization and Experiment Tracking*, in detail):

```
%reload_ext tensorboard
%tensorboard --logdir logs
```

17. Now, we will use the model for inference to check if it works properly. Let's define a prediction function to simplify the prediction steps, as follows:

```
def get_prediction(text):
    inputs = tokenizer(text, padding=True,truncation=True,
    max_length=250, return_tensors="pt").to(device)
    outputs = \ model(inputs["input_ids"].
to(device),inputs["attention_mask"].to(device))
    probs = outputs[0].softmax(1)
    return probs, probs.argmax()
```

18. Now, run the model for inference:

```
>>> text = "I didn't like the movie it bored me "
>>> get_prediction(text)[1].item()
0
```

19. What we got here is 0, which is a negative. We have already defined which ID refers to which label. We can use this mapping scheme to get the label. Alternatively, we can simply pass all these boring steps to a dedicated API, namely Pipeline, which we are already familiar with. Before instantiating it, let's save the best model for further inference:

```
model_save_path = "MyBestIMDBModel"
trainer.save_model(model_save_path)
tokenizer.save_pre-trained(model_save_path)
```

The Pipeline API is an easy way to use pre-trained models for inference. We load the model from where we saved it and pass it to the Pipeline API, which does the rest. We can skip this saving step and instead directly pass our `model` and `tokenizer` objects in memory to the Pipeline API. If you do so, you will get the same result.

20. As shown in the following code, we need to specify the task name argument of Pipeline as `sentiment-analysis` when we perform binary classification:

```
>>> from transformers import pipeline,
\ DistilBertForSequenceClassification,
DistilBertTokenizerFast
>>> model = \ DistilBertForSequenceClassification.from_
pre-trained("MyBestIMDBModel")
>>> tokenizer= \ DistilBertTokenizerFast.from_
pre-trained("MyBestIMDBModel")
>>> nlp= pipeline("sentiment-analysis", model=model,
tokenizer=tokenizer)
>>> nlp("the movie was very impressive")
Out:  [{'label': 'POS', 'score': 0.9621992707252502}]
>>> nlp("the text of the picture was very poor")
Out:  [{'label': 'NEG', 'score': 0.9938313961029053}]
```

Pipeline knows how to treat the input and somehow learned which ID refers to which (POS or NEG) label. It also yields the class probabilities.

Well done! We have fine-tuned a sentiment prediction model for the IMDb dataset using the `Trainer` class. In the next section, we will do the same binary classification training but with native PyTorch. We will also use a different dataset.

Training a classification model with native PyTorch

The `Trainer` class is very powerful, and we have the HuggingFace team to thank for providing such a useful tool. However, in this section, we will fine-tune the pre-trained model from scratch to see what happens under the hood. Let's get started:

1. First, let's load the model for fine-tuning. We will select `DistilBERT` here since it is a small, fast, and cheap version of BERT:

    ```
    from transformers import
    DistilBertForSequenceClassification
    model = DistilBertForSequenceClassification.from_
    pre-trained('distilbert-base-uncased')
    ```

2. To fine-tune any model, we need to put it into training mode, as follows:

    ```
    model.train()
    ```

3. Now, we must load the tokenizer:

    ```
    from transformers import DistilBertTokenizerFast
    tokenizer = DistilBertTokenizerFast.from_
    pre-trained('bert-base-uncased')
    ```

4. Since the `Trainer` class organized the entire process for us, we did not deal with optimization and other training settings in the previous IMDb sentiment classification exercise. Now, we need to instantiate the optimizer ourselves. Here, we must select `AdamW`, which is an implementation of the Adam algorithm but with a weight decay fix. Recently, it has been shown that `AdamW` produces better training loss and validation loss than models trained with Adam. Hence, it is a widely used optimizer within many transformer training processes:

    ```
    from transformers import AdamW
    optimizer = AdamW(model.parameters(), lr=1e-3)
    ```

To design the fine-tuning process from scratch, we must understand how to implement a single step forward and backpropagation. We can pass a single batch through the transformer layer and get the output, which is called **forward propagation**. Then, we must compute the loss using the output and ground truth label and update the model weight based on the loss. This is called **backpropagation**.

The following code receives three sentences associated with the labels in a single batch and performs forward propagation. At the end, the model automatically computes the loss:

```
import torch
texts= ["this is a good example","this is a bad
example","this is a good one"]
labels= [1,0,1]
labels = torch.tensor(labels).unsqueeze(0)
encoding = tokenizer(texts, return_tensors='pt',
padding=True,
truncation=True, max_length=512)
input_ids = encoding['input_ids']
attention_mask = encoding['attention_mask']
outputs = \
model(input_ids, attention_mask=attention_mask,
labels=labels)
loss = outputs.loss
loss.backward()
optimizer.step()
Outputs
SequenceClassifierOutput(
[('loss', tensor(0.7178, grad_fn=<NllLossBackward>)),
('logits',tensor([[ 0.0664, -0.0161],[ 0.0738, 0.0665], [
0.0690, -0.0010]], grad_fn=<AddmmBackward>))])
```

The model takes `input_ids` and `attention_mask`, which were produced by the tokenizer, and computes the loss using ground truth labels. As we can see, the output consists of both `loss` and `logits`. Now, `loss.backward()` computes the gradient of the tensor by evaluating the model with the inputs and labels. `optimizer.step()` performs a single optimization step and updates the weight using the gradients that were computed, which is called backpropagation. When we put all these lines into a loop shortly, we will also add `optimizer.zero_grad()`, which clears the gradient of all the parameters. It is important to call this at the beginning of the loop; otherwise, we may accumulate the gradients from multiple steps. The second tensor of the output is **logits**. In the context of deep learning, the term logits (short for **logistic units**) is the last layer of the neural architecture and consists of prediction values as real numbers. Logits need to be turned into probabilities by the softmax function in the case of classification. Otherwise, they are simply normalized for regression.

5. If we want to manually calculate the loss, we must not pass the labels to the model. Due to this, the model only yields the logits and does not calculate the loss. In the following example, we are computing the cross-entropy loss manually:

```
from torch.nn import functional
labels = torch.tensor([1,0,1])
outputs = model(input_ids, attention_mask=attention_mask)
loss = functional.cross_entropy(outputs.logits, labels)
loss.backward()
optimizer.step()
loss
Output: tensor(0.6101, grad_fn=<NllLossBackward>)
```

6. With that, we've learned how batch input is fed in the forward direction through the network in a single step. Now, it is time to design a loop that iterates over the entire dataset in batches to train the model with several epochs. To do so, we will start by designing the `Dataset` class. It is a subclass of `torch.Dataset`, inherits member variables and functions, and implements `__init__()` and `__getitem()__` abstract functions:

```
from torch.utils.data import Dataset
class MyDataset(Dataset):
    def __init__(self, encodings, labels):
        self.encodings = encodings
        self.labels = labels
```

```
        def __getitem__(self, idx):
            item = {key: torch.tensor(val[idx]) for key, val
    in self.encodings.items()}
            item['labels'] = torch.tensor(self.labels[idx])
            return item
        def __len__(self):
            return len(self.labels)
```

7. Let's fine-tune the model for sentiment analysis by taking another sentiment
    analysis dataset called the SST-2 dataset; that is, **Stanford Sentiment Treebank v2**
    (**SST2**). We will also load the corresponding metric for SST-2 for evaluation,
    as follows:

```
import datasets
from datasets import load_dataset
sst2= load_dataset("glue","sst2")
from datasets import load_metric
metric = load_metric("glue", "sst2")
```

8. We will extract the sentences and the labels accordingly:

```
texts=sst2['train']['sentence']
labels=sst2['train']['label']
val_texts=sst2['validation']['sentence']
val_labels=sst2['validation']['label']
```

9. Now, we can pass the datasets through the tokenizer and instantiate the
    MyDataset object to make the BERT models work with them:

```
train_dataset= MyDataset(tokenizer(texts,
truncation=True, padding=True), labels)
val_dataset=  MyDataset(tokenizer(val_texts,
truncation=True, padding=True), val_labels)
```

10. Let's instantiate a `Dataloader` class that provides an interface to iterate through the data samples by loading order. This also helps with batching and memory pinning:

```
from torch.utils.data import DataLoader
train_loader = DataLoader(train_dataset, batch_size=16,
shuffle=True)
val_loader  =  DataLoader(val_dataset, batch_size=16,
shuffle=True)
```

11. The following lines detect the device and define the `AdamW` optimizer properly:

```
from transformers import  AdamW
device = \
torch.device('cuda') if torch.cuda.is_available() else
torch.device('cpu')
model.to(device)
optimizer = AdamW(model.parameters(), lr=1e-3)
```

So far, we know how to implement forward propagation, which is where we process a batch of examples. Here, batch data is fed in the forward direction through the neural network. In a single step, each layer from the first to the final one is processed by the batch data, as per the activation function, and is passed to the successive layer. To go through the entire dataset in several epochs, we designed two nested loops: the outer loop is for the epoch, while the inner loop is for the steps for each batch. The inner part is made up of two blocks; one is for training, while the other one is for evaluating each epoch. As you may have noticed, we called `model.train()` at the first training loop, and when we moved the second evaluation block, we called `model.eval()`. This is important as we put the model into training and inference mode.

12. We have already discussed the inner block. Note that we track the model's performance by means of the corresponding the `metric` object:

```
for epoch in range(3):
    model.train()
    for batch in train_loader:
        optimizer.zero_grad()
        input_ids = batch['input_ids'].to(device)
        attention_mask = batch['attention_mask'].
to(device)
```

```
        labels = batch['labels'].to(device)
        outputs = \
model(input_ids, attention_mask=attention_mask,
labels=labels)
            loss = outputs[0]
            loss.backward()
            optimizer.step()
        model.eval()
        for batch in val_loader:
            input_ids = batch['input_ids'].to(device)
            attention_mask = batch['attention_mask'].
to(device)
            labels = batch['labels'].to(device)
            outputs = \
model(input_ids, attention_mask=attention_mask,
labels=labels)
            predictions=outputs.logits.argmax(dim=-1)
            metric.add_batch(
                    predictions=predictions,
                    references=batch["labels"],
                )
        eval_metric = metric.compute()
        print(f"epoch {epoch}: {eval_metric}")
OUTPUT:
epoch 0: {'accuracy': 0.9048165137614679}
epoch 1: {'accuracy': 0.8944954128440367}
epoch 2: {'accuracy': 0.9094036697247706}
```

Well done! We've fine-tuned our model and got around 90.94 accuracy. The remaining processes, such as saving, loading, and inference, will be similar to what we did with the `Trainer` class.

With that, we are done with binary classification. In the next section, we will learn how to implement a model for multi-class classification for a language other than English.

Fine-tuning BERT for multi-class classification with custom datasets

In this section, we will fine-tune the Turkish BERT, namely **BERTurk**, to perform seven-class classification downstream tasks with a custom dataset. This dataset has been compiled from Turkish newspapers and consists of seven categories. We will start by getting the dataset. Alternatively, you can find it in this book's GitHub respository or get it from `https://www.kaggle.com/savasy/ttc4900`:

1. First, run the following code to get data within a Python notebook:

    ```
    !wget https://raw.githubusercontent.com/savasy/
    TurkishTextClassification/master/TTC4900.csv
    ```

2. Start by loading the data:

    ```
    import pandas as pd
    data= pd.read_csv("TTC4900.csv")
    data=data.sample(frac=1.0, random_state=42)
    ```

3. Let's organize the IDs and labels with `id2label` and `label2id` to make the model figure out which ID refers to which label. We will also pass the number of labels, `NUM_LABELS`, to the model to specify the size of a thin classification head layer on top of the BERT model:

    ```
    labels=["teknoloji","ekonomi","saglik","siyaset",
    "kultur","spor","dunya"]
    NUM_LABELS= len(labels)
    id2label={i:l for i,l in enumerate(labels)}
    label2id={l:i for i,l in enumerate(labels)}
    data["labels"]=data.category.map(lambda x: label2id[x.
    strip()])
    data.head()
    ```

The output is as follows:

	category	text	labels
4657	teknoloji	acıların kedisi sam çatık kaşlı kedi sam in i...	0
3539	spor	g saray a git santos van_persie den forma ala...	5
907	dunya	endonezya da çatışmalar 14 ölü endonezya da i...	6
4353	teknoloji	emniyetten polis logolu virüs uyarısı telefon...	0
3745	spor	beni türk yapın cristian_baroni yıldırım dan ...	5

Figure 5.5 – Text classification dataset – TTC 4900

4. Let's count and plot the number of classes using a pandas object:

```
data.category.value_counts().plot(kind='pie')
```

As shown in the following diagram, the dataset classes have been fairly distributed:

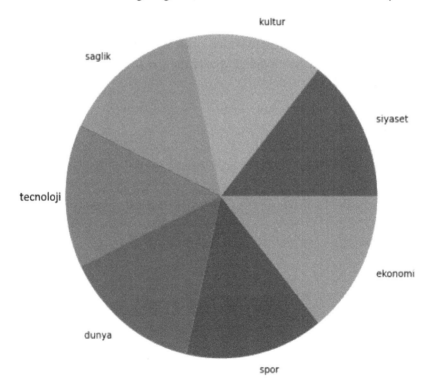

Figure 5.6 – The class distribution

5. The following execution instantiates a sequence classification model with the number of labels (7), label ID mappings, and a Turkish BERT model (dbmdz/bert-base-turkish-uncased), namely BERTurk. To check this, execute the following:

```
>>> model
```

6. The output will be a summary of the model and is too long to show here. Instead, let's turn our attention to the last layer by using the following code:

```
(classifier): Linear(in_features=768, out_features=7,
bias=True)
```

7. You may have noticed that we did not choose DistilBert as there is no pre-trained *uncased* DistilBert for the Turkish language:

```
from transformers import BertTokenizerFast
tokenizer = BertTokenizerFast.from_pre-trained("dbmdz/
bert-base-turkish-uncased", max_length=512)
  from transformers import BertForSequenceClassification
model = BertForSequenceClassification.from_
pre-trained("dbmdz/bert-base-turkish-uncased", num_
labels=NUM_LABELS, id2label=id2label, label2id=label2id)
model.to(device)
```

8. Now, let's prepare the training (%50), validation (%25), and test (%25) datasets, as follows:

```
SIZE= data.shape[0]
## sentences
train_texts= list(data.text[:SIZE//2])
val_texts=   list(data.text[SIZE//2:(3*SIZE)//4 ])
test_texts=  list(data.text[(3*SIZE)//4:])
## labels
train_labels= list(data.labels[:SIZE//2])
val_labels=   list(data.labels[SIZE//2:(3*SIZE)//4])
test_labels=  list(data.labels[(3*SIZE)//4:])
## check the size
len(train_texts), len(val_texts), len(test_texts)
(2450, 1225, 1225)
```

9. The following code tokenizes the sentences of three datasets and their tokens and converts them into integers (`input_ids`), which are then fed into the BERT model:

```
train_encodings = tokenizer(train_texts, truncation=True,
padding=True)
val_encodings  = tokenizer(val_texts, truncation=True,
padding=True)
test_encodings = tokenizer(test_texts, truncation=True,
padding=True)
```

10. We have already implemented the `MyDataset` class (please see page 14). The class inherits from the abstract `Dataset` class by overwriting the __getitem__ and __len__() methods, which are expected to return the items and the size of the dataset using any data loader, respectively:

```
train_dataset = MyDataset(train_encodings, train_labels)
val_dataset = MyDataset(val_encodings, val_labels)
test_dataset = MyDataset(test_encodings, test_labels)
```

11. We will keep batch size as 16 since we have a relatively small dataset. Notice that the other parameters of `TrainingArguments` are almost the same as they were for the previous sentiment analysis experiment:

```
from transformers import TrainingArguments, Trainer
training_args = TrainingArguments(
    output_dir='./TTC4900Model',
    do_train=True,
    do_eval=True,
    num_train_epochs=3,
    per_device_train_batch_size=16,
    per_device_eval_batch_size=32,
    warmup_steps=100,
    weight_decay=0.01,
    logging_strategy='steps',
    logging_dir='./multi-class-logs',
    logging_steps=50,
    evaluation_strategy="steps",
    eval_steps=50,
```

```
        save_strategy="epoch",
        fp16=True,
        load_best_model_at_end=True
)
```

12. Sentiment analysis and text classification are objects of the same evaluation metrics; that is, macro-averaging macro-averaged F1, Precision, and Recall. Therefore, we will not define the `compute_metric()` function again. Here is the code for instantiating a `Trainer` object:

```
trainer = Trainer(
    model=model,
    args=training_args,
    train_dataset=train_dataset,
    eval_dataset=val_dataset,
    compute_metrics= compute_metrics
)
```

13. Finally, let's start the training process:

```
trainer.train()
```

The output is as follows:

[462/462 12:27, Epoch 3/3]

Step	Training Loss	Validation Loss	Accuracy	F1	Precision	Recall	Runtime	Samples Per Second
50	1.874100	1.706715	0.377143	0.379416	0.553955	0.383715	20.982000	58.383000
100	0.842900	0.327575	0.915102	0.913738	0.914565	0.915279	20.981700	58.384000
150	0.358200	0.281808	0.911020	0.910288	0.912012	0.911213	20.997000	58.342000
200	0.233500	0.366845	0.905306	0.905313	0.916948	0.903440	20.980800	58.387000
250	0.222700	0.292270	0.922449	0.921374	0.921567	0.923131	20.981300	58.385000
300	0.257700	0.280120	0.924898	0.923810	0.924427	0.924510	20.979800	58.390000
350	0.115200	0.292410	0.925714	0.924946	0.924752	0.925454	20.982700	58.381000
400	0.064900	0.322697	0.925714	0.924944	0.925674	0.925265	20.988300	58.366000
450	0.080400	0.297606	0.929796	0.929267	0.929170	0.929497	20.985100	58.375000

The minimum loss

Figure 5.7 – The output of the Trainer class for text classification

14. To check the trained model, we must evaluate the fine-tuned model on three dataset splits, as follows. Our best model is fine-tuned at step 300 with a loss of 0.28012:

```
q=[trainer.evaluate(eval_dataset=data) for data in
[train_dataset, val_dataset, test_dataset]]
pd.DataFrame(q, index=["train","val","test"]).iloc[:,:5]
```

The output is as follows:

[77/77 01:24]

	eval_loss	eval_Accuracy	eval_F1	eval_Precision	eval_Recall
train	0.091844	0.975510	0.97546	0.975942	0.975535
val	0.280120	0.924898	0.92381	0.924427	0.924510
test	0.280038	0.926531	0.92542	0.927410	0.925425

Figure 5.8 – The text classification model's performance on the train/validation/test dataset

The classification accuracy is around 92.6, while the F1 macro-average is around 92.5. In the literature, many approaches have been tested on this Turkish benchmark dataset. They mostly followed TF-IDF and linear classifier, word2vec embeddings, or an LSTM-based classifier and got around 90.0 F1 at best. Compared to those approaches, other than transformer, the fine-tuned BERT model outperforms them.

15. As with any other experiment, we can track the experiment via TensorBoard:

```
%load_ext tensorboard
%tensorboard --logdir multi-class-logs/
```

16. Let's design a function that will run the model for inference. If you want to see a real label instead of an ID, you can use the config object of our model, as shown in the following predict function:

```
def predict(text):
    inputs = tokenizer(text, padding=True,
truncation=True, max_length=512, return_tensors="pt").
to("cuda")
    outputs = model(**inputs)
    probs = outputs[0].softmax(1)
    return probs, probs.argmax(),model.config.
id2label[probs.argmax().item()]
```

17. Now, we are ready to call the predict function for text classification inference. The following code classifies a sentence about a football team:

```
text = "Fenerbahçeli futbolcular kısa paslarla hazırlık
çalışması yaptılar"
predict(text)
(tensor([[5.6183e-04, 4.9046e-04, 5.1385e-04, 9.9414e-04,
3.4417e-04, 9.9669e-01, 4.0617e-04]], device='cuda:0',
grad_fn=<SoftmaxBackward>), tensor(5, device='cuda:0'),
'spor')
```

18. As we can see, the model correctly predicted the sentence as sports (`spor`). Now, it is time to save the model and reload it using the `from_pre-trained()` function. Here is the code:

```
model_path = "turkish-text-classification-model"
trainer.save_model(model_path)
tokenizer.save_pre-trained(model_path)
```

19. Now, we can reload the saved model and run inference with the help of the `pipeline` class:

```
model_path = "turkish-text-classification-model"
from transformers import pipeline,
BertForSequenceClassification, BertTokenizerFast
model = BertForSequenceClassification.from_
pre-trained(model_path)
tokenizer= BertTokenizerFast.from_pre-trained(model_path)
nlp= pipeline("sentiment-analysis", model=model,
tokenizer=tokenizer)
```

20. You may have noticed that the task's name is `sentiment-analysis`. This term may be confusing but this argument will actually return `TextClassificationPipeline` at the end. Let's run the pipeline:

```
>>> nlp("Sinemada hangi filmler oynuyor bugün")
[{'label': 'kultur', 'score': 0.9930670261383057}]
>>> nlp("Dolar ve Euro bugün yurtiçi piyasalarda
yükseldi")
[{'label': 'ekonomi', 'score': 0.9927696585655212}]
>>> nlp("Bayern Münih ile Barcelona bugün karşı karşıya
geliyor. Maçı İngiliz hakem James Watts yönetecek!")
[{'label': 'spor', 'score': 0.9975664019584656}]
```

That's our model! It has predicted successfully.

So far, we have implemented two single-sentence tasks; that is, sentiment analysis and multi-class classification. In the next section, we will learn how to handle sentence-pair input and how to design a regression model with BERT.

Fine-tuning the BERT model for sentence-pair regression

The regression model is considered to be for classification, but the last layer only contains a single unit. This is not processed by softmax logistic regression but normalized. To specify the model and put a single-unit head layer at the top, we can either directly pass the num_labels=1 parameter to the BERT.from_pre-trained() method or pass this information through a Config object. Initially, this needs to be copied from the config object of the pre-trained model, as follows:

```
from transformers import DistilBertConfig,
DistilBertTokenizerFast, DistilBertForSequenceClassification
model_path='distilbert-base-uncased'
config = DistilBertConfig.from_pre-trained(model_path, num_
labels=1)
tokenizer = DistilBertTokenizerFast.from_pre-trained(model_
path)
model = \
DistilBertForSequenceClassification.from_pre-trained(model_
path, config=config)
```

Well, our pre-trained model has a single-unit head layer thanks to the num_labels=1 parameter. Now, we are ready to fine-tune the model with our dataset. Here, we will use the **Semantic Textual Similarity-Benchmark (STS-B)**, which is a collection of sentence pairs that have been drawn from a variety of content, such as news headlines. Each pair has been annotated with a similarity score from 1 to 5. Our task is to fine-tune the BERT model to predict these scores. We will evaluate the model using the Pearson/Spearman correlation coefficients while following the literature. Let's get started:

1. The following code loads the data. The original data was splits into three. However, the test split has no label so that we can divide the validation data into two parts, as follows:

```
import datasets
from datasets import load_dataset
stsb_train= load_dataset('glue','stsb', split="train")
```

```
stsb_validation = load_dataset('glue','stsb',
split="validation")
stsb_validation=stsb_validation.shuffle(seed=42)
stsb_val= datasets.Dataset.from_dict(stsb_
validation[:750])
stsb_test= datasets.Dataset.from_dict(stsb_
validation[750:])
```

2. Let's make the `stsb_train` training data neat by wrapping it with pandas:

```
pd.DataFrame(stsb_train)
```

Here is what the training data looks like:

	idx	label	sentence1	sentence2
0	0	5.00	A plane is taking off.	An air plane is taking off.
1	1	3.80	A man is playing a large flute.	A man is playing a flute.
2	2	3.80	A man is spreading shreded cheese on a pizza.	A man is spreading shredded cheese on an uncoo...
3	3	2.60	Three men are playing chess.	Two men are playing chess.
4	4	4.25	A man is playing the cello.	A man seated is playing the cello.
...	...	...	...	...
5744	5744	0.00	Severe Gales As Storm Clodagh Hits Britain	Merkel pledges NATO solidarity with Latvia
5745	5745	0.00	Dozens of Egyptians hostages taken by Libyan t...	Egyptian boat crash death toll rises as more b...
5746	5746	0.00	President heading to Bahrain	President Xi: China to continue help to fight ...
5747	5747	0.00	China, India vow to further bilateral ties	China Scrambles to Reassure Jittery Stock Traders
5748	5748	0.00	Putin spokesman: Doping charges appear unfounded	The Latest on Severe Weather: 1 Dead in Texas ...

5749 rows × 4 columns

Figure 5.9 – STS-B training dataset

3. Run the following code to check the shape of the three sets:

```
stsb_train.shape, stsb_val.shape, stsb_test.shape
((5749, 4), (750, 4), (750, 4))
```

4. Run the following code to tokenize the datasets:

```
enc_train = stsb_train.map(lambda e: tokenizer(
e['sentence1'],e['sentence2'], padding=True,
truncation=True), batched=True, batch_size=1000)
enc_val =    stsb_val.map(lambda e: tokenizer(
e['sentence1'],e['sentence2'], padding=True,
truncation=True), batched=True, batch_size=1000)
enc_test =   stsb_test.map(lambda e: tokenizer(
e['sentence1'],e['sentence2'], padding=True,
truncation=True), batched=True, batch_size=1000)
```

5. The tokenizer merges two sentences with a `[SEP]` delimiter and produces single `input_ids` and an `attention_mask` for a sentence pair, as shown here:

```
pd.DataFrame(enc_train)
```

The output is as follows:

	attention_mask	idx	input_ids	label	sentence1	sentence2
0	[1, 1, 1, 1, 1, 1, 1, 1, 1, 1, 1, 1, 1, 1,...	0	[101, 1037, 4946, 2003, 2635, 2125, 1012, 102,...	5.00	A plane is taking off.	An air plane is taking off.
1	[1, 1, 1, 1, 1, 1, 1, 1, 1, 1, 1, 1, 1, 1,...	1	[101, 1037, 2158, 2003, 2652, 1037, 2312, 8928...	3.80	A man is playing a large flute.	A man is playing a flute.
2	[1, 1, 1, 1, 1, 1, 1, 1, 1, 1, 1, 1, 1, 1,...	2	[101, 1037, 2158, 2003, 9359, 14021, 5596, 209...	3.80	A man is spreading shreded cheese on a pizza.	A man is spreading shredded cheese on an uncoo...
3	[1, 1, 1, 1, 1, 1, 1, 1, 1, 1, 1, 1, 1, 1,...	3	[101, 2093, 2273, 2024, 2652, 7433, 1012, 102,...	2.60	Three men are playing chess.	Two men are playing chess.
4	[1, 1, 1, 1, 1, 1, 1, 1, 1, 1, 1, 1, 1, 1,...	4	[101, 1037, 2158, 2003, 2652, 1996, 10145, 101...	4.25	A man is playing the cello.	A man seated is playing the cello.
...	...	...	...	...	...	...
5744	[1, 1, 1, 1, 1, 1, 1, 1, 1, 1, 1, 1, 1, 1,...	5744	[101, 5729, 14554, 2015, 2004, 4040, 18856, 13...	0.00	Severe Gales As Storm Clodagh Hits Britain	Merkel pledges NATO solidarity with Latvia
5745	[1, 1, 1, 1, 1, 1, 1, 1, 1, 1, 1, 1, 1, 1,...	5745	[101, 9877, 1997, 23437, 19323, 2579, 2011, 19...	0.00	Dozens of Egyptians hostages taken by Libyan t...	Egyptian boat crash death toll rises as more b...
5746	[1, 1, 1, 1, 1, 1, 1, 1, 1, 1, 1, 1, 1, 1,...	5746	[101, 2343, 5825, 2000, 15195, 102, 2343, 8418...	0.00	President heading to Bahrain	President Xi: China to continue help to fight ...
5747	[1, 1, 1, 1, 1, 1, 1, 1, 1, 1, 1, 1, 1, 1,...	5747	[101, 2859, 1010, 2634, 19076, 2000, 2582, 177...	0.00	China, India vow to further bilateral ties	China Scrambles to Reassure Jittery Stock Traders
5748	[1, 1, 1, 1, 1, 1, 1, 1, 1, 1, 1, 1, 1, 1,...	5748	[101, 22072, 14056, 1024, 23799, 5571, 3711, 4...	0.00	Putin spokesman: Doping charges appear unfounded	The Latest on Severe Weather: 1 Dead in Texas ...

5749 rows × 6 columns

Figure 5.10 – Encoded training dataset

Similar to other experiments, we follow almost the same scheme for the `TrainingArguments` and `Trainer` classes. Here is the code:

```
from transformers import TrainingArguments, Trainer
training_args = TrainingArguments(
    output_dir='./stsb-model',
    do_train=True,
    do_eval=True,
    num_train_epochs=3,
    per_device_train_batch_size=32,
    per_device_eval_batch_size=64,
    warmup_steps=100,
    weight_decay=0.01,
```

```
        logging_strategy='steps',
        logging_dir='./logs',
        logging_steps=50,
        evaluation_strategy="steps",
        save_strategy="epoch",
        fp16=True,
        load_best_model_at_end=True
    )
```

6. Another important difference between the current regression task and the previous classification tasks is the design of `compute_metrics`. Here, our evaluation metric will be based on the **Pearson Correlation Coefficient** and the **Spearman's Rank Correlation** following the common practice provided in the literature. We also provide the **Mean Squared Error** (**MSE**), **Root Mean Square Error** (**RMSE**), and **Mean Absolute Error** (**MAE**) metrics, which are commonly used, especially for regression models:

```
import numpy as np
from scipy.stats import pearsonr
from scipy.stats import spearmanr
def compute_metrics(pred):
    preds = np.squeeze(pred.predictions)
    return {"MSE": ((preds - pred.label_ids) **
2).mean().item(),
            "RMSE": (np.sqrt((  (preds - pred.label_ids)
** 2).mean())).item(),
            "MAE": (np.abs(preds - pred.label_ids)).
mean().item(),
        "Pearson" : pearsonr(preds,pred.label_ids)[0],
        "Spearman's Rank":spearmanr(preds,pred.label_ids)[0]
            }
```

7. Now, let's instantiate the `Trainer` object:

```
trainer = Trainer(
        model=model,
        args=training_args,
        train_dataset=enc_train,
        eval_dataset=enc_val,
```

```
                          compute_metrics=compute_metrics,
                          tokenizer=tokenizer
          )
```

Run the training, like so:

```
          train_result = trainer.train()
```

The output is as follows:

Step	Training Loss	Validation Loss	Mse	Rmse	Mae	Pearson	Spearman's rank	Runtime	Samples Per Second
50	4.973200	2.242550	2.242550	1.497515	1.261815	0.140489	0.138228	0.943900	794.538000
100	1.447300	0.801587	0.801587	0.895314	0.735321	0.808588	0.809430	0.933300	803.602000
150	0.940400	0.693730	0.693730	0.832904	0.675787	0.843234	0.842421	0.930700	805.838000
200	0.736300	0.679696	0.679696	0.824437	0.662136	0.846722	0.843393	0.934700	802.407000
250	0.585400	0.590002	0.590002	0.768116	0.618677	0.859470	0.854824	0.931600	805.067000
300	0.513800	0.584674	0.584674	0.764640	0.610141	0.861033	0.856779	0.942900	795.438000
350	0.488000	0.604512	0.604512	0.777504	0.611338	0.865844	0.861726	0.939600	798.174000
400	0.362900	0.555219	0.555219	0.745130	0.582900	0.868366	0.863372	0.938200	799.379000
450	0.298500	0.544973	0.544973	0.738223	0.576751	0.868145	0.864209	0.938200	799.407000
500	0.270100	0.546966	0.546966	0.739571	0.575326	0.867538	0.864035	0.941900	796.240000

Figure 5.11 – Training result for text regression

8. The best validation loss that's computed is 0.544973 at step 450. Let's evaluate the best checkpoint model at that step, as follows:

```
          q=[trainer.evaluate(eval_dataset=data) for data in [enc_
          train, enc_val, enc_test]]
          pd.DataFrame(q, index=["train","val","test"]).iloc[:,:5]
```

The output is as follows:

[90/90 00:30]

	eval_loss	eval_MSE	eval_RMSE	eval_MAE	eval_Pearson	eval_Spearman's Rank
train	0.232471	0.232471	0.482152	0.372915	0.944844	0.935578
val	0.544973	0.544973	0.738223	0.576751	0.868145	0.864209
test	0.537752	0.537752	0.733316	0.567489	0.875409	0.872858

Figure 5.12 – Regression performance on the training/validation/test dataset

The Pearson and Spearman correlation scores are around 87.54 and 87.28 on the test dataset, respectively. We did not get a SoTA result, but we did get a comparable result for the STS-B task based on the GLUE Benchmark leaderboard. Please check the leaderboard!

9. We are now ready to run the model for inference. Let's take the following two sentences, which share the same meaning, and pass them to the model:

```
s1,s2="A plane is taking off.","An air plane is taking
off."
encoding = tokenizer(s1,s2, return_tensors='pt',
padding=True, truncation=True, max_length=512)
input_ids = encoding['input_ids'].to(device)
attention_mask = encoding['attention_mask'].to(device)
outputs = model(input_ids, attention_mask=attention_mask)
outputs.logits.item()
OUTPUT: 4.033723831176758
```

10. The following code consumes the negative sentence pair, which means the sentences are semantically different:

```
s1,s2="The men are playing soccer.","A man is riding a
motorcycle."
encoding = tokenizer("hey how are you there","hey how are
you", return_tensors='pt', padding=True, truncation=True,
max_length=512)
input_ids = encoding['input_ids'].to(device)
attention_mask = encoding['attention_mask'].to(device)
outputs = model(input_ids, attention_mask=attention_mask)
outputs.logits.item()
OUTPUT: 2.3579328060150146
```

11. Finally, we will save the model, as follows:

```
model_path = "sentence-pair-regression-model"
trainer.save_model(model_path)
tokenizer.save_pre-trained(model_path)
```

Well done! We can congratulate ourselves since we have successfully completed three tasks: sentiment analysis, multi-class classification, and sentence pair regression.

Utilizing run_glue.py to fine-tune the models

So far, we have designed a fine-tuning architecture from scratch using both native PyTorch and the `Trainer` class. The HuggingFace community also provides another powerful script called `run_glue.py` for GLUE benchmark and GLUE-like classification downstream tasks. This script can handle and organize the entire training/validation process for us. If you want to do quick prototyping, you should use this script. It can fine-tune any pre-trained models on the HuggingFace hub. We can also feed it with our own data in any format.

Please go to the following link to access the script and to learn more: `https://github.com/huggingface/transformers/tree/master/examples`.

The script can perform nine different GLUE tasks. With the script, we can do everything that we have done with the `Trainer` class so far. The task name could be one of the following GLUE tasks: `cola`, `sst2`, `mrpc`, `stsb`, `qqp`, `mnli`, `qnli`, `rte`, or `wnli`.

Here is the script scheme for fine-tuning a model:

```
export TASK_NAME= "My-Task-Name"
python run_glue.py \
  --model_name_or_path bert-base-cased \
  --task_name $TASK_NAME \
  --do_train \   --do_eval \
  --max_seq_length 128 \
  --per_device_train_batch_size 32 \
  --learning_rate 2e-5 \
  --num_train_epochs 3 \
  --output_dir /tmp/$TASK_NAME/
```

The community provides another script called `run_glue_no_trainer.py`. The main difference between the original script and this one is that this no-trainer script gives us more chances to change the options for the optimizer, or add any customization that we want to do.

Summary

In this chapter, we discussed how to fine-tune a pre-trained model for any text classification downstream task. We fine-tuned the models using sentiment analysis, multi-class classification, and sentence-pair classification – more specifically, sentence-pair regression. We worked with a well-known IMDb dataset and our own custom dataset to train the models. While we took advantage of the `Trainer` class to cope with much of the complexity of the processes for training and fine-tuning, we learned how to train from scratch with native libraries to understand forward propagation and backpropagation with the `transformers` library. To summarize, we discussed and conducted fine-tuning single-sentence classification with Trainer, sentiment classification with native PyTorch without Trainer, single-sentence multi-class classification, and fine-tuning sentence-pair regression.

In the next chapter, we will learn how to fine-tune a pre-trained model to any token classification downstream task, such as parts-of-speech tagging or named-entity recognition.

6
Fine-Tuning Language Models for Token Classification

In this chapter, we will learn about fine-tuning language models for token classification. Tasks such as **Named Entity Recognition (NER)**, **Part-of-Speech (POS)** tagging, and **Question Answering (QA)** are explored in this chapter. We will learn how a specific language model can be fine-tuned on such tasks. We will focus on BERT more than other language models. You will learn how to apply POS, NER, and QA using BERT. You will get familiar with the theoretical details of these tasks such as their respective datasets and how to perform them. After finishing this chapter, you will be able to perform any token classification using Transformers.

In this chapter, we will fine-tune BERT for the following tasks: fine-tuning BERT for token classification problems such as NER and POS, fine-tuning a language model for an NER problem, and thinking of the QA problem as a start/stop token classification.

The following topics will be covered in this chapter:

- Introduction to token classification
- Fine-tuning language models for NER
- Question answering using token classification

Technical requirements

We will be using Jupyter Notebook to run our coding exercises and Python 3.6+ and the following packages need to be installed:

- `sklearn`
- `transformers 4.0+`
- `Datasets`
- `seqeval`

All notebooks with coding exercises will be available at the following GitHub link: `https://github.com/PacktPublishing/Mastering-Transformers/tree/main/CH06`.

Check out the following link to see the Code in Action video: `https://bit.ly/2UGMQP2`

Introduction to token classification

The task of classifying each token in a token sequence is called **token classification**. This task says that a specific model must be able to classify each token into a class. POS and NER are two of the most well-known tasks in this criterion. However, QA is also another major NLP task that fits in this category. We will discuss the basics of these three tasks in the following sections.

Understanding NER

One of the well-known tasks in the category of token classification is NER – the recognition of each token as an entity or not and identifying the type of each detected entity. For example, a text can contain multiple entities at the same time – person names, locations, organizations, and other types of entities. The following text is a clear example of NER:

George Washington is one the presidents of the United States of America.

George Washington is a person name while *the United States of America* is a location name. A sequence tagging model is expected to tag each word in the form of tags, each containing information about the tag. BIO's tags are the ones that are universally used for standard NER tasks.

The following table is a list of tags and their descriptions:

Tag	Description
O	Out of entity
B-PER	Beginning of Person entity
I-PER	Inside of Person entity
B-LOC	Beginning of Location entity
I-LOC	Inside of Location entity
B-ORG	Beginning of Organization entity
I-ORG	Inside of Organization entity
B-MISC	Beginning of Miscellaneous entity
I-MISC	Inside of Miscellaneous entity

Table 1 – Table of BIOS tags and their descriptions

From this table, **B** indicates the beginning of a tag, and **I** denotes the inside of a tag, while **O** is the outside of the entity. This is the reason that this type of annotation is called **BIO**. For example, the sentence shown earlier can be annotated using BIO:

```
[B-PER|George] [I-PER|Washington] [O|is] [O|one] [O|the]
[O|presidents] [O|of] [B-LOC|United] [I-LOC|States] [I-LOC|of]
[I-LOC|America] [O|.]
```

Accordingly, the sequence must be tagged in BIO format. A sample dataset can be in the format shown as follows:

```
SOCCER NN B-NP O
- : O O
JAPAN NNP B-NP B-LOC
GET VB B-VP O
LUCKY NNP B-NP O
WIN NNP I-NP O
, , O O
CHINA NNP B-NP B-PER
IN IN B-PP O
SURPRISE DT B-NP O
DEFEAT NN I-NP O
. . O O

Nadim NNP B-NP B-PER
Ladki NNP I-NP I-PER

AL-AIN NNP B-NP B-LOC
, , O O
United NNP B-NP B-LOC
Arab NNP I-NP I-LOC
Emirates NNPS I-NP I-LOC
1996-12-06 CD I-NP O
```

Figure 6.1 – CONLL2003 dataset

In addition to the NER tags we have seen, there are POS tags available in this dataset

Understanding POS tagging

POS tagging, or grammar tagging, is annotating a word in a given text according to its respective part of speech. As a simple example, in a given text, identification of each word's role in the categories of noun, adjective, adverb, and verb is considered to be POS. However, from a linguistic perspective, there are many roles other than these four.

In the case of POS tags, there are variations, but the Penn Treebank POS tagset is one of the most well-known ones. The following screenshot shows a summary and respective description of these roles:

1. CC	Coordinating conjunction	25. TO	*to*
2. CD	Cardinal number	26. UH	Interjection
3. DT	Determiner	27. VB	Verb, base form
4. EX	Existential *there*	28. VBD	Verb, past tense
5. FW	Foreign word	29. VBG	Verb, gerund/present
6. IN	Preposition/subordinating		participle
	conjunction	30. VBN	Verb, past participle
7. JJ	Adjective	31. VBP	Verb, non-3rd ps. sing. present
8. JJR	Adjective, comparative	32. VBZ	Verb, 3rd ps. sing. present
9. JJS	Adjective, superlative	33. WDT	*wh*-determiner
10. LS	List item marker	34. WP	*wh*-pronoun
11. MD	Modal	35. WP$	Possessive *wh*-pronoun
12. NN	Noun, singular or mass	36. WRB	*wh*-adverb
13. NNS	Noun, plural	37. #	Pound sign
14. NNP	Proper noun, singular	38. $	Dollar sign
15. NNPS	Proper noun, plural	39. .	Sentence-final punctuation
16. PDT	Predeterminer	40. ,	Comma
17. POS	Possessive ending	41. :	Colon, semi-colon
18. PRP	Personal pronoun	42. (	Left bracket character
19. PP$	Possessive pronoun	43.)	Right bracket character
20. RB	Adverb	44. "	Straight double quote
21. RBR	Adverb, comparative	45. '	Left open single quote
22. RBS	Adverb, superlative	46. "	Left open double quote
23. RP	Particle	47. '	Right close single quote
24. SYM	Symbol (mathematical or scientific)	48. "	Right close double quote

Figure 6.2 – Penn Treebank POS tags

Datasets for POS tasks are annotated like the example shown in *Figure 6.1*.

The annotation of these tags is very useful in specific NLP applications and is one of the building blocks of many other methods. Transformers and many advanced models can somehow understand the relation of words in their complex architecture.

Understanding QA

A QA or reading comprehension task comprises a set of reading comprehension texts with respective questions on them. An exemplary dataset from this scope is **SQUAD** or **Stanford Question Answering Dataset**. This dataset consists of Wikipedia texts and respective questions asked about them. The answers are in the form of segments of the original Wikipedia text.

The following screenshot shows an example of this dataset:

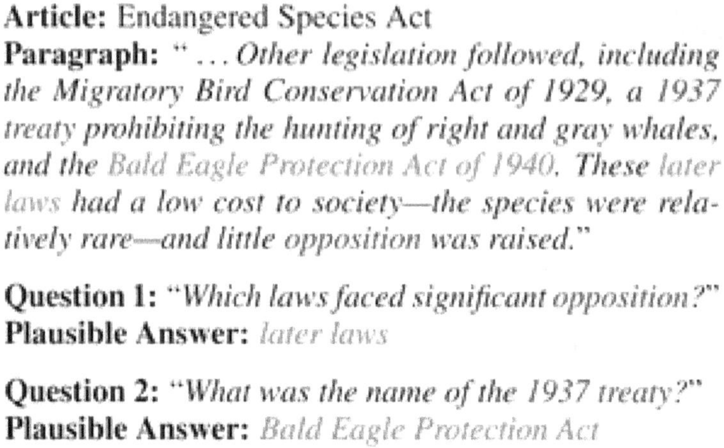

Article: Endangered Species Act

Paragraph: " ... *Other legislation followed, including the Migratory Bird Conservation Act of 1929, a 1937 treaty prohibiting the hunting of right and gray whales, and the* Bald Eagle Protection Act of 1940. **These** later laws *had a low cost to society—the species were relatively rare—and little opposition was raised.*"

Question 1: *"Which laws faced significant opposition?"*
Plausible Answer: later laws

Question 2: *"What was the name of the 1937 treaty?"*
Plausible Answer: Bald Eagle Protection Act

Figure 6.3 – SQUAD dataset example

The highlighted red segments are the answers and important parts of each question are highlighted in blue. It is required for a good NLP model to segment text according to the question, and this segmentation can be done in the form of sequence labeling. The model labels the start and the end of the segment as answer start and end segments.

Up to this point, you have learned the basics of modern NLP sequence tagging tasks such as QA, NER, and POS. In the next section, you will learn how it is possible to fine-tune BERT for these specific tasks and use the related datasets from the `datasets` library.

Fine-tuning language models for NER

In this section, we will learn how to fine-tune BERT for an NER task. We first start with the `datasets` library and by loading the `conll2003` dataset.

The dataset card is accessible at `https://huggingface.co/datasets/conll2003`. The following screenshot shows this model card from the HuggingFace website:

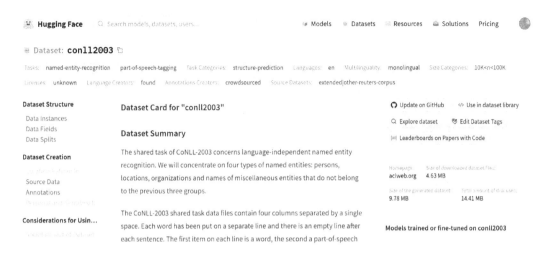

Figure 6.4 – CONLL2003 dataset card from HuggingFace

From this screenshot, it can be seen that the model is trained on this dataset and is currently available and listed in the right panel. However, there are also descriptions of the dataset such as its size and its characteristics:

1. To load the dataset, the following commands are used:

```
import datasets
conll2003 = datasets.load_dataset("conll2003")
```

A download progress bar will appear and after finishing the downloading and caching, the dataset will be ready to use. The following screenshot shows the progress bars:

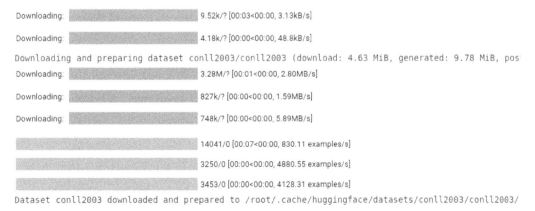

Figure 6.5 – Downloading and preparing the dataset

2. You can easily double-check the dataset by accessing the train samples using the following command:

```
>>> conll2003["train"][0]
```

The following screenshot shows the result:

```
{'chunk_tags': [11, 21, 11, 12, 21, 22, 11, 12, 0],
 'id': '0',
 'ner_tags': [3, 0, 7, 0, 0, 0, 7, 0, 0],
 'pos_tags': [22, 42, 16, 21, 35, 37, 16, 21, 7],
 'tokens': ['EU',
  'rejects',
  'German',
  'call',
  'to',
  'boycott',
  'British',
  'lamb',
  '.']}
```

Figure 6.6 – CONLL2003 train samples from the datasets library

3. The respective tags for POS and NER are shown in the preceding screenshot. We will use only NER tags for this part. You can use the following command to get the NER tags available in this dataset:

```
>>> conll2003["train"].features["ner_tags"]
```

4. The result is also shown in *Figure 6.7*. All the BIO tags are shown and there are nine tags in total:

```
>>> Sequence(feature=ClassLabel(num_classes=9,
names=['O', 'B-PER', 'I-PER', 'B-ORG', 'I-ORG', 'B-LOC',
'I-LOC', 'B-MISC', 'I-MISC'], names_file=None, id=None),
length=-1, id=None)
```

5. The next step is to load the BERT tokenizer:

```
from transformers import BertTokenizerFast
tokenizer = BertTokenizerFast.from_pretrained("bert-base-
uncased")
```

6. The `tokenizer` class can work with white-space tokenized sentences also. We need to enable our tokenizer for working with white-space tokenized sentences, because the NER task has a token-based label for each token. Tokens in this task are usually the white-space tokenized words rather than BPE or any other tokenizer tokens. According to what is said, let's see how `tokenizer` can be used with a white-space tokenized sentence:

```
>>>
tokenizer(["Oh","this","sentence","is","tokenized","and",
"splitted","by","spaces"], is_split_into_words=True)
```

As you can see, by just setting `is_split_into_words` to `True`, the problem is solved.

7. It is required to preprocess the data before using it for training. To do so, we must use the following function and map into the entire dataset:

```
def tokenize_and_align_labels(examples):
    tokenized_inputs = tokenizer(examples["tokens"],
            truncation=True, is_split_into_words=True)
    labels = []
    for i, label in enumerate(examples["ner_tags"]):
        word_ids = \
         tokenized_inputs.word_ids(batch_index=i)
        previous_word_idx = None
        label_ids = []
        for word_idx in word_ids:
            if word_idx is None:
                label_ids.append(-100)
            elif word_idx != previous_word_idx:
                label_ids.append(label[word_idx])
            else:
                label_ids.append(label[word_idx] if
  label_all_tokens else -100)
            previous_word_idx = word_idx
        labels.append(label_ids)
    tokenized_inputs["labels"] = labels
    return tokenized_inputs
```

8. This function will make sure that our tokens and labels are aligned properly. This alignment is required because the tokens are tokenized in pieces, but the words must be of one piece. To test and see how this function works, you can run it by giving a single sample to it:

```
q = tokenize_and_align_labels(conll2003['train'][4:5])
print(q)
```

And the result is shown as follows:

```
>>> {'input_ids': [[101, 2762, 1005, 1055, 4387, 2000,
1996, 2647, 2586, 1005, 1055, 15651, 2837, 14121, 1062,
9328, 5804, 2056, 2006, 9317, 10390, 2323, 4965, 8351,
4168, 4017, 2013, 3032, 2060, 2084, 3725, 2127, 1996,
4045, 6040, 2001, 24509, 1012, 102]], 'token_type_ids':
[[0, 0, 0, 0, 0, 0, 0, 0, 0, 0, 0, 0, 0, 0, 0, 0, 0, 0,
0, 0, 0, 0, 0, 0, 0, 0, 0, 0, 0, 0, 0, 0, 0, 0, 0, 0, 0,
0, 0]], 'attention_mask': [[1, 1, 1, 1, 1, 1, 1, 1, 1, 1,
1, 1, 1, 1, 1, 1, 1, 1, 1, 1, 1, 1, 1, 1, 1, 1, 1, 1, 1,
1, 1, 1, 1, 1, 1, 1, 1, 1, 1]], 'labels': [[-100, 5, 0,
-100, 0, 0, 0, 3, 4, 0, -100, 0, 0, 1, 2, -100, -100, 0,
0, 0, 0, 0, 0, -100, -100, 0, 0, 0, 0, 5, 0, 0, 0, 0,
0, 0, 0, -100]]}
```

9. But this result is not readable, so you can run the following code to have a readable version:

```
for token, label in zip(tokenizer.convert_ids_to_
tokens(q["input_ids"][0]),q["labels"][0]):
    print(f"{token:_<40} {label}")
```

The result is shown as follows:

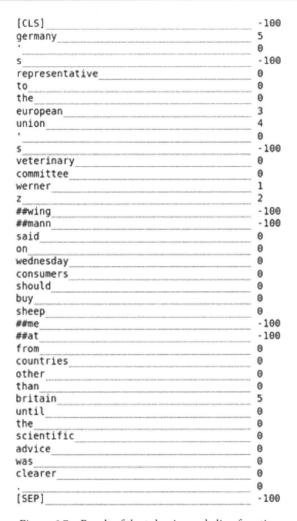

[CLS]	-100
germany	5
'	0
s	-100
representative	0
to	0
the	0
european	3
union	4
'	0
s	-100
veterinary	0
committee	0
werner	1
z	2
##wing	-100
##mann	-100
said	0
on	0
wednesday	0
consumers	0
should	0
buy	0
sheep	0
##me	-100
##at	-100
from	0
countries	0
other	0
than	0
britain	5
until	0
the	0
scientific	0
advice	0
was	0
clearer	0
.	0
[SEP]	-100

Figure 6.7 – Result of the tokenize and align functions

10. The mapping of this function to the dataset can be done by using the map function of the datasets library:

```
>>> tokenized_datasets = \ conll2003.map(tokenize_and_
align_labels, batched=True)
```

11. In the next step, it is required to load the BERT model with the respective number of labels:

```
from transformers import\ AutoModelForTokenClassification
model = AutoModelForTokenClassification.from_
pretrained("bert-base-uncased", num_labels=9)
```

12. The model will be loaded and ready to be trained. In the next step, we must prepare the trainer and training parameters:

```
from transformers import TrainingArguments, Trainer
args = TrainingArguments(
"test-ner",
evaluation_strategy = "epoch",
learning_rate=2e-5,
per_device_train_batch_size=16,
per_device_eval_batch_size=16,
num_train_epochs=3,
weight_decay=0.01,
)
```

13. It is required to prepare the data collator. It will apply batch operations on the training dataset to use less memory and perform faster. You can do so as follows:

```
from transformers import \
DataCollatorForTokenClassification
data_collator = \
DataCollatorForTokenClassification(tokenizer)
```

14. To be able to evaluate model performance, there are many metrics available for many tasks in HuggingFace's datasets library. We will be using the sequence evaluation metric for NER. seqeval is a good Python framework to evaluate sequence tagging algorithms and models. It is necessary to install the seqeval library:

```
pip install seqeval
```

15. Afterward, you can load the metric:

```
>>> metric = datasets.load_metric("seqeval")
```

16. It is easily possible to see how the metric works by using the following code:

```
example = conll2003['train'][0]
label_list = \ conll2003["train"].features["ner_tags"].
feature.names
labels = [label_list[i] for i in example["ner_tags"]]
metric.compute(predictions=[labels], references=[labels])
```

The result is as follows:

```
{'MISC': {'f1': 1.0, 'number': 2, 'precision': 1.0, 'recall': 1.0},
 'ORG': {'f1': 1.0, 'number': 1, 'precision': 1.0, 'recall': 1.0},
 'overall_accuracy': 1.0,
 'overall_f1': 1.0,
 'overall_precision': 1.0,
 'overall_recall': 1.0}
```

Figure 6.8 – Output of the seqeval metric

Various metrics such as accuracy, F1-score, precision, and recall are computed for the sample input.

17. The following function is used to compute the metrics:

```
import numpy as np def compute_metrics(p):
    predictions, labels = p
    predictions = np.argmax(predictions, axis=2)
    true_predictions = [
        [label_list[p] for (p, 1) in zip(prediction,
label) if 1 != -100]
        for prediction, label in zip(predictions,
labels)    ]
    true_labels = [
        [label_list[1] for (p, 1) in zip(prediction, label)
if 1 != -100]
        for prediction, label in zip(predictions, labels)
    ]
    results = \
        metric.compute(predictions=true_predictions,
        references=true_labels)
    return {
    "precision": results["overall_precision"],
    "recall": results["overall_recall"],
    "f1": results["overall_f1"],
    "accuracy": results["overall_accuracy"],
    }
```

18. The last steps are to make a trainer and train it accordingly:

```
trainer = Trainer(
    model,
    args,
```

```
        train_dataset=tokenized_datasets["train"],
        eval_dataset=tokenized_datasets["validation"],
        data_collator=data_collator,
        tokenizer=tokenizer,
        compute_metrics=compute_metrics
    )
    trainer.train()
```

19. After running the `train` function of `trainer`, the result will be as follows:

[2634/2634 20:09, Epoch 3/3]

Epoch	Training Loss	Validation Loss	Precision	Recall	F1	Accuracy	Runtime	Samples Per Second
1	0.035800	0.043440	0.937072	0.944800	0.940920	0.988454	17.061700	190.486000
2	0.019100	0.043591	0.939359	0.951531	0.945406	0.989311	16.797100	193.486000
3	0.014500	0.043591	0.939359	0.951531	0.945406	0.989311	16.790100	193.567000

Figure 6.9 – Trainer results after running train

20. It is necessary to save the model and tokenizer after training:

```
model.save_pretrained("ner_model")
tokenizer.save_pretrained("tokenizer")
```

21. If you wish to use the model with the pipeline, you must read the config file and assign `label2id` and `id2label` correctly according to the labels you have used in the `label_list` object:

```
id2label = {
str(i): label for i,label in enumerate(label_list)
}
label2id = {
label: str(i) for i,label in enumerate(label_list)
}
import json
config = json.load(open("ner_model/config.json"))
config["id2label"] = id2label
config["label2id"] = label2id
json.dump(config, open("ner_model/config.json","w"))
```

22. Afterward, it is easy to use the model as in the following example:

```
from transformers import pipeline
model = \ AutoModelForTokenClassification.from_
pretrained("ner_model")
nlp = \
pipeline("ner", model=mmodel, tokenizer=tokenizer)
example = "I live in Istanbul"
ner_results = nlp(example)
print(ner_results)
```

And the result will appear as seen here:

```
[{'entity': 'B-LOC', 'score': 0.9983942, 'index': 4,
'word': 'istanbul', 'start': 10, 'end': 18}]
```

Up to this point, you have learned how to apply POS using BERT. You learned how to train your own POS tagging model using Transformers and you also tested the model. In the next section, we will focus on QA.

Question answering using token classification

A **QA** problem is generally defined as an NLP problem with a given text and a question for AI, and getting an answer back. Usually, this answer can be found in the original text but there are different approaches to this problem. In the case of **Visual Question Answering (VQA)**, the question is about a visual entity or visual concept rather than text but the question itself is in the form of text.

Some examples of VQA are as follows:

Figure 6.10 – VQA examples

Most of the models that are intended to be used in VQA are multimodal models that can understand the visual context along with the question and generate the answer properly. However, unimodal fully textual QA or just QA is based on textual context and textual questions with respective textual answers:

1. SQUAD is one of the most well-known datasets in the field of QA. To see examples of SQUAD and examine them, you can use the following code:

```
from pprint import pprint
from datasets import load_dataset
squad = load_dataset("squad")
for item in squad["train"][1].items():
    print(item[0])
    pprint(item[1])
    print("="*20)
```

The following is the result:

```
answers
{'answer_start': [188], 'text': ['a copper statue of
Christ']}

====================
Context
('Architecturally, the school has a Catholic character.
Atop the Main ' "Building's gold dome is a golden statue
of the Virgin Mary. Immediately in " 'front of the Main
Building and facing it, is a copper statue of Christ with
' 'arms upraised with the legend "Venite Ad Me Omnes".
Next to the Main ' 'Building is the Basilica of the
Sacred Heart. Immediately behind the ' 'basilica is the
Grotto, a Marian place of prayer and reflection. It is
a ' 'replica of the grotto at Lourdes, France where the
Virgin Mary reputedly ' 'appeared to Saint Bernadette
Soubirous in 1858. At the end of the main drive ' '(and
in a direct line that connects through 3 statues and
the Gold Dome), is ' 'a simple, modern stone statue of
Mary.')

====================
Id
'5733be284776f4190066117f'

====================
Question
```

```
'What is in front of the Notre Dame Main Building?'
=====================
Title
'University_of_Notre_Dame'
=====================
```

However, there is version 2 of the SQUAD dataset, which has more training samples, and it is highly recommended to use it. To have an overall understanding of how it is possible to train a model for a QA problem, we will focus on the current part of this problem.

2. To start, load SQUAD version 2 using the following code:

```
from datasets import load_dataset
squad = load_dataset("squad_v2")
```

3. After loading the SQUAD dataset, you can see the details of this dataset by using the following code:

```
>>> squad
```

The result is as follows:

```
DatasetDict({
    train: Dataset({
        features: ['id', 'title', 'context', 'question', 'answers'],
        num_rows: 130319
    })
    validation: Dataset({
        features: ['id', 'title', 'context', 'question', 'answers'],
        num_rows: 11873
    })
})
```

Figure 6.11 – SQUAD dataset (version 2) details

The details of the SQUAD dataset will be shown as seen in *Figure 6.11*. As you can see, there are more than 130,000 training samples with more than 11,000 validation samples.

4. As we did for NER, we must preprocess the data to have the right form to be used by the model. To do so, you must first load your tokenizer, which is a pretrained tokenizer as long as you are using a pretrained model and want to fine-tune it for a QA problem:

```
from transformers import AutoTokenizer
model = "distilbert-base-uncased"
tokenizer = AutoTokenizer.from_pretrained(model)
```

As you have seen, we are going to use the distillBERT model.

According to our SQUAD example, we need to give more than one text to the model, one for the question and one for the context. Accordingly, we need our tokenizer to put these two side by side and separate them with the special [SEP] token because distillBERT is a BERT-based model.

There is another problem in the scope of QA, and it is the size of the context. The context size can be longer than the model input size, but we cannot reduce it to the size the model accepts. With some problems, we can do so but in QA, it is possible that the answer could be in the truncated part. We will show you an example where we tackle this problem using document stride.

5. The following is an example to show how it works using tokenizer:

```
max_length = 384
doc_stride = 128
example = squad["train"][173]
tokenized_example = tokenizer(
example["question"],
example["context"],
max_length=max_length,
truncation="only_second",
return_overflowing_tokens=True,
stride=doc_stride
)
```

6. The stride is the document stride used to return the stride for the second part, like a window, while the return_overflowing_tokens flag gives the model information on whether it should return the extra tokens. The result of tokenized_example is more than a single tokenized output, instead having two input IDs. In the following, you can see the result:

```
>>> len(tokenized_example['input_ids'])
>>> 2
```

7. Accordingly, you can see the full result by running the following for loop:

```
for input_ids in tokenized_example["input_ids"][:2]:
    print(tokenizer.decode(input_ids))
    print("-"*50)
```

The result is as follows:

[CLS] beyonce got married in 2008 to whom? [SEP] on april
4, 2008, beyonce married jay z. she publicly revealed
their marriage in a video montage at the listening party
for her third studio album, i am... sasha fierce, in
manhattan's sony club on october 22, 2008. i am... sasha
fierce was released on november 18, 2008 in the united
states. the album formally introduces beyonce's alter
ego sasha fierce, conceived during the making of her
2003 single " crazy in love ", selling 482, 000 copies
in its first week, debuting atop the billboard 200, and
giving beyonce her third consecutive number - one album
in the us. the album featured the number - one song "
single ladies (put a ring on it) " and the top - five
songs " if i were a boy " and " halo ". achieving the
accomplishment of becoming her longest - running hot
100 single in her career, " halo "'s success in the us
helped beyonce attain more top - ten singles on the list
than any other woman during the 2000s. it also included
the successful " sweet dreams ", and singles " diva ",
" ego ", " broken - hearted girl " and " video phone ".
the music video for " single ladies " has been parodied
and imitated around the world, spawning the " first
major dance craze " of the internet age according to the
toronto star. the video has won several awards, including
best video at the 2009 mtv europe music awards, the 2009
scottish mobo awards, and the 2009 bet awards. at the
2009 mtv video music awards, the video was nominated for
nine awards, ultimately winning three including video
of the year. its failure to win the best female video
category, which went to american country pop singer
taylor swift's " you belong with me ", led to kanye west
interrupting the ceremony and beyonce [SEP]

[CLS] beyonce got married in 2008 to whom? [SEP] single
ladies " has been parodied and imitated around the world,
spawning the " first major dance craze " of the internet
age according to the toronto star. the video has won
several awards, including best video at the 2009 mtv
europe music awards, the 2009 scottish mobo awards, and
the 2009 bet awards. at the 2009 mtv video music awards,
the video was nominated for nine awards, ultimately
winning three including video of the year. its failure
to win the best female video category, which went to

```
american country pop singer taylor swift's " you belong
with me ", led to kanye west interrupting the ceremony
and beyonce improvising a re - presentation of swift's
award during her own acceptance speech. in march 2009,
beyonce embarked on the i am... world tour, her second
headlining worldwide concert tour, consisting of 108
shows, grossing $ 119. 5 million. [SEP]

----------------------------------------------------
```

As you can see from the preceding output, with a window of 128 tokens, the rest of the context is replicated again in the second output of input IDs.

Another problem is the end span, which is not available in the dataset, but instead, the start span or the start character for the answer is given. It is easy to find the length of the answer and add it to the start span, which would automatically yield the end span.

8. Now that we know all the details of this dataset and how to deal with them, we can easily put them together to make a preprocessing function (link: https://github.com/huggingface/transformers/blob/master/examples/pytorch/question-answering/run_qa.py):

```python
def prepare_train_features(examples):
    # tokenize examples
    tokenized_examples = tokenizer(
        examples["question" if pad_on_right else
"context"],
        examples["context" if pad_on_right else
"question"],
        truncation="only_second" if pad_on_right else
"only_first",
        max_length=max_length,
        stride=doc_stride,
        return_overflowing_tokens=True,
        return_offsets_mapping=True,
        padding="max_length",
    )
    # map from a feature to its example
    sample_mapping = \ tokenized_examples.pop("overflow_
to_sample_mapping")
    offset_mapping = \ tokenized_examples.pop("offset_
```

```
mapping")
    tokenized_examples["start_positions"] = []
    tokenized_examples["end_positions"] = []
    # label impossible answers with CLS
    # start and end token are the answers for each one
    for i, offsets in enumerate(offset_mapping):
        input_ids = tokenized_examples["input_ids"][i]
        cls_index = \ input_ids.index(tokenizer.cls_
token_id)
        sequence_ids = \ tokenized_examples.sequence_
ids(i)
        sample_index = sample_mapping[i]
        answers = examples["answers"][sample_index]
        if len(answers["answer_start"]) == 0:
            tokenized_examples["start_positions"].\
append(cls_index)
            tokenized_examples["end_positions"].\
append(cls_index)
        else:
            start_char = answers["answer_start"][0]
            end_char = \
                start_char + len(answers["text"][0])
            token_start_index = 0
            while sequence_ids[token_start_index] != / (1
if pad_on_right else 0):
                token_start_index += 1
            token_end_index = len(input_ids) - 1
            while sequence_ids[token_end_index] != (1 if
pad_on_right else 0):
                token_end_index -= 1
            if not (offsets[token_start_index][0] <=
start_char and offsets[token_end_index][1] >= end_char):
                tokenized_examples["start_positions"].
append(cls_index)
                tokenized_examples["end_positions"].
append(cls_index)
            else:
                while token_start_index < len(offsets)
```

```
and offsets[token_start_index][0] <= start_char:
                token_start_index += 1
            tokenized_examples["start_positions"].
append(token_start_index - 1)
                while offsets[token_end_index][1] >= end_
char:
                token_end_index -= 1
            tokenized_examples["end_positions"].
append(token_end_index + 1)
    return tokenized_examples
```

9. Mapping this function to the dataset would apply all the required changes:

```
>>> tokenized_datasets = squad.map(prepare_train_
features, batched=True, remove_columns=squad["train"].
column_names)
```

10. Just like other examples, you can now load a pretrained model to be fine-tuned:

```
from transformers import AutoModelForQuestionAnswering,
TrainingArguments, Trainer

model = AutoModelForQuestionAnswering.from_
pretrained(model)
```

11. The next step is to create training arguments:

```
args = TrainingArguments(
"test-squad",
evaluation_strategy = "epoch",
learning_rate=2e-5,
per_device_train_batch_size=16,
per_device_eval_batch_size=16,
num_train_epochs=3,
weight_decay=0.01,
)
```

12. If we are not going to use a data collator, we will give a default data collator to the model trainer:

```
from transformers import default_data_collator
data_collator = default_data_collator
```

13. Now, everything is ready to make the trainer:

```
trainer = Trainer(
model,
args,
train_dataset=tokenized_datasets["train"],
eval_dataset=tokenized_datasets["validation"],
data_collator=data_collator,
tokenizer=tokenizer,
)
```

14. And the trainer can be used with the `train` function:

```
trainer.train()
```

The result will be something like the following:

Epoch	Training Loss	Validation Loss	Runtime	Samples Per Second
1	1.220600	1.160322	39.574900	272.496000
2	0.945200	1.121690	39.706000	271.596000
3	0.773000	1.157358	39.734000	271.405000

[16599/16599 58:06, Epoch 3/3]

Figure 6.12 – Training results

As you can see, the model is trained with three epochs and the outputs for loss in validation and training are reported.

15. Like any other model, you can easily save this model by using the following function:

```
>>> trainer.save_model("distillBERT_SQUAD")
```

If you want to use your saved model or any other model that is trained on QA, the `transformers` library provides a pipeline that's easy to use and implement with no extra effort.

16. By using this pipeline functionality, you can use any model. The following is an example given for using a model with the QA pipeline:

```
from transformers import pipeline
qa_model = pipeline('question-answering',
model='distilbert-base-cased-distilled-squad',
tokenizer='distilbert-base-cased')
```

The pipeline just requires two inputs to make the model ready for usage, the model and the tokenizer. Although, you are also required to give it a pipeline type, which is QA in the given example.

17. The next step is to give it the inputs it requires, `context` and `question`:

```
>>> question = squad["validation"][0]["question"]
>>> context = squad["validation"][0]["context"]
The question and the context can be seen by using
following code:
>>> print("Question:")
>>> print(question)
>>> print("Context:")
>>> print(context)
Question:
In what country is Normandy located?
Context:
('The Normans (Norman: Nourmands; French: Normands;
Latin: Normanni) were the ' 'people who in the 10th and
11th centuries gave their name to Normandy, a ' 'region
in France. They were descended from Norse ("Norman"
comes from ' '"Norseman") raiders and pirates from
Denmark, Iceland and Norway who, under ' 'their leader
Rollo, agreed to swear fealty to King Charles III of
West ' 'Francia. Through generations of assimilation
and mixing with the native ' 'Frankish and Roman-Gaulish
populations, their descendants would gradually ' 'merge
with the Carolingian-based cultures of West Francia. The
distinct ' 'cultural and ethnic identity of the Normans
emerged initially in the first ' 'half of the 10th
century, and it continued to evolve over the succeeding '
'centuries.')
```

18. The model can be used by the following example:

```
>>> qa_model(question=question, context=context)
```

And the result can be seen as follows:

```
{'answer': 'France', 'score': 0.9889379143714905,
'start': 159, 'end': 165,}
```

Up to this point, you have learned how you can train on the dataset you want. You have also learned how you can use the trained model using pipelines.

Summary

In this chapter, we discussed how to fine-tune a pretrained model to any token classification task. Fine-tuning models on NER and QA problems were explored. Using the pretrained and fine-tuned models on specific tasks with pipelines was detailed with examples. We also learned about various preprocessing steps for these two tasks. Saving pretrained models that are fine-tuned on specific tasks was another major learning point of this chapter. We also saw how it is possible to train models with a limited input size on tasks such as QA that have longer sequence sizes than the model input. Using tokenizers more efficiently to have document splitting with document stride was another important item in this chapter too.

In the next chapter, we will discuss text representation methods using Transformers. By studying the chapter, you will learn how to perform zero-/few-shot learning and semantic text clustering.

7
Text Representation

So far, we have addressed classification and generation problems with the `transformers` library. Text representation is another crucial task in modern **Natural Language Processing** (**NLP**), especially for unsupervised tasks such as clustering, semantic search, and topic modeling. Representing sentences by using various models such as **Universal Sentence Encoder** (**USE**) and Siamese BERT (Sentence-BERT) with additional libraries such as sentence transformers will be explained here. Zero-shot learning using BART will also be explained, and you will learn how to utilize it. Few-shot learning methodologies and unsupervised use cases such as semantic text clustering and topic modeling will also be described. Finally, one-shot learning use cases such as semantic search will be covered.

The following topics will be covered in this chapter:

- Introduction to sentence embeddings
- Benchmarking sentence similarity models
- Using BART for zero-shot learning
- Semantic similarity experiment with FLAIR
- Text clustering with Sentence-BERT
- Semantic search with Sentence-BERT

Technical requirements

We will be using a Jupyter notebook to run our coding exercises. For this, you will need Python 3.6+ and the following packages:

- `sklearn`

- `transformers >=4.00`

- `datasets`

- `sentence-transformers`

- `tensorflow-hub`

- `flair`

- `umap-learn`

- `bertopic`

All the notebooks for the coding exercises in this chapter will be available at the following GitHub link:

`https://github.com/PacktPublishing/Mastering-Transformers/tree/main/CH07`

Check out the following link to see the Code in Action video:
`https://bit.ly/2VcMCyI`

Introduction to sentence embeddings

Pre-trained BERT models do not produce efficient and independent sentence embeddings as they always need to be fine-tuned in an end-to-end supervised setting. This is because we can think of a pre-trained BERT model as an indivisible whole and semantics is spread across all layers, not just the final layer. Without fine-tuning, it may be ineffective to use its internal representations independently. It is also hard to handle unsupervised tasks such as clustering, topic modeling, information retrieval, or semantic search. Because we have to evaluate many sentence pairs during clustering tasks, for instance, this causes massive computational overhead.

Luckily, many modifications have been made to the original BERT model, such as **Sentence-BERT** (**SBERT**), to derive semantically meaningful and independent sentence embeddings. We will talk about these approaches in a moment. In the NLP literature, many neural sentence embedding methods have been proposed for mapping a single sentence to a common feature space (vector space model) wherein a cosine function (or dot product) is usually used to measure similarity and the Euclidean distance to measure dissimilarity.

The following are some applications that can be efficiently solved with sentence embeddings:

- Sentence-pair tasks
- Information retrieval
- Question answering
- Duplicate question detection
- Paraphrase detection
- Document clustering
- Topic modeling

The simplest but most efficient kind of neural sentence embedding is the average-pooling operation, which is performed on the embeddings of words in a sentence. To get a better representation of this, some early neural methods learned sentence embeddings in an unsupervised fashion, such as Doc2Vec, Skip-Thought, FastSent, and Sent2Vec. Doc2Vec utilized a token-level distributional theory and an objective function to predict adjacent words, similar to Word2Vec. The approach injects an additional memory token (called **Paragraph-ID**) into each sentence, which is reminiscent of CLS or SEP tokens in the `transformers` library. This additional token acts as a piece of memory that represents the context or document embeddings. SkipThought and FastSent are considered sentence-level approaches, where the objective function is used to predict adjacent sentences. These models extract the sentence's meaning to obtain necessary information from the adjacent sentences and their context.

Some other methods, such as InferSent, leveraged supervised learning and multi-task transfer learning to learn generic sentence embeddings. InferSent trained various supervised tasks to get more efficient embedding. RNN-based supervised models such as GRU or LSTM utilize the last hidden state (or stacked entire hidden states) to obtain sentence embeddings in a supervised setting. We touched on the RNN approach in *Chapter 1, From Bag-of-Words to the Transformers.*

Cross-encoder versus bi-encoder

So far, we have discussed how to train Transformer-based language models and fine-tune them in semi-supervised and supervised settings, respectively. As we learned in the previous chapters, we got successful results thanks to the transformer architectures. Once a task-specific thin linear layer has been put on top of a pre-trained model, all the weights of the network (not only the last task-specific thin layer) are fine-tuned with task-specific labeled data. We also experienced how the BERT architecture has been fine-tuned for two different groups of tasks (single-sentence or sentence-pair) without any architectural modifications being required. The only difference is that for the sentence-pair tasks, the sentences are concatenated and marked with a SEP token. Thus, self-attention is applied to all the tokens of the concatenated sentences. This is a big advantage of the BERT model, where both input sentences can get the necessary information from each other at every layer. In the end, they are encoded simultaneously. This is called cross-encoding.

However, there are two disadvantages regarding the cross-encoders that were addressed by the SBERT authors and *Humeau et al., 2019*, as follows:

- The cross-encoder setup is not convenient for many sentence-pair tasks due to too many possible combinations needing to be processed. For instance, to get the two closest sentences from a list of 1,000 sentences, the cross-encoder model (BERT) requires around 500,000 ($n * (n-1) /2$) inference computation. Therefore, it would be very slow compared to alternative solutions such as SBERT or USE. This is because these alternatives produce independent sentence embeddings wherein the similarity function (cosine similarity) or dissimilarity function (Euclidean or Manhattan) can easily be applied. Note that these dis/similarity functions can be performed efficiently on modern architectures. Moreover, with the help of an optimized index structure, we can reduce computational complexity from many hours to a few minutes when comparing or clustering many documents.

- Due to its supervised characteristics, the BERT model can't derive independent meaningful sentence embeddings. It is hard to leverage a pre-trained BERT model as is for unsupervised tasks such as clustering, semantic search, or topic modeling. The BERT model produces a fixed-size vector for each token in a document. In an unsupervised setting, the document-level representation may be obtained by averaging or pooling token vectors, plus SEP and CLS tokens. Later, we will see that such a representation of BERT produces below-average sentence embeddings, and that its performance scores are usually worse than word embedding pooling techniques such as Word2Vec, FastText, or GloVe.

Alternatively, bi-encoders (such as SBERT) independently map a sentence pair to a semantic vector space, as shown in the following diagram. Since the representations are separate, bi-encoders can cache the encoded input representation for each input, resulting in fast inference time. One of the successful bi-encoder modifications of BERT is SBERT. Based on the Siamese and Triplet network structures, SBERT fine-tunes the BERT model to produce semantically meaningful and independent embeddings of the sentences.

The following diagram shows the bi-encoder architecture:

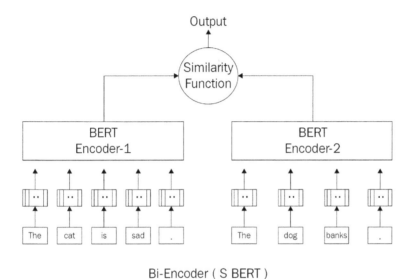

Figure 7.1 – Bi-encoder architecture

You can find hundreds of pre-trained SBERT models that have been trained with different objectives at `https://public.ukp.informatik.tu-darmstadt.de/reimers/sentence-transformers/v0.2/`.

We will use some of them in the next section.

Benchmarking sentence similarity models

There are many *semantic textual similarity* models available, but it is highly recommended that you benchmark and understand their capabilities and differences using metrics. *Papers With Code* provides a list of these datasets at `https://paperswithcode.com/task/semantic-textual-similarity`.

Also, there are many model outputs in each dataset that are ranked by their results. These results have been taken from the aforementioned article.

GLUE provides most of these datasets and tests, but it is not only for semantic textual similarity. **GLUE**, which stands for **General Language Understanding Evaluation**, is a general benchmark for evaluating a model with different NLP characteristics. More details about the GLUE dataset and its usage was provided in *Chapter 2, A Hands-On Introduction to the Subject*. Let's take a look at it before we move on:

1. To load the metrics and MRPC dataset from the GLUE benchmark, you can use the following code:

    ```
    from datasets import load_metric, load_dataset
    metric = load_metric('glue', 'mrpc')
    mrpc = load_dataset('glue', 'mrpc')
    ```

    The samples in this dataset are labeled 1 and 0, which indicates whether they are similar or dissimilar, respectively. You can use any model, regardless of the architecture, to produce values for two given sentences. In other words, the model should classify the two sentences as zeros and ones.

2. Let's assume the model produces values and that these values are stored in an array called `predictions`. You can easily use this metric with the predictions to see the F1 and accuracy values:

    ```
    labels = [i['label'] for i in dataset['test']]
    metric.compute(predictions=predictions,
    references=labels)
    ```

3. Some semantic textual similarity datasets such as **Semantic Textual Similarity Benchmark (STSB)** have different metrics. For example, this benchmark uses Spearman and Pearson correlations because the outputs and predictions are between 0 and 5 and are float numbers instead of being 0s and 1s, which is a regression problem. The following code shows an example of this benchmark:

    ```
    metric = load_metric('glue', 'stsb')
    metric.compute(predictions=[1,2,3],references=[5,2,2])
    ```

    The predictions and references are the same as the ones from the **Microsoft Research Paraphrase Corpus (MRPC)**; the predictions are the model outputs, while the references are the dataset labels.

4. To get a comparative result between two models, we will use a distilled version of Roberta and test these two on the STSB. To start, you must load both models. The following code shows how to install the required libraries before loading and using models:

```
pip install tensorflow-hub
pip install sentence-transformers
```

5. As we mentioned previously, the next step is to load the dataset and metric:

```
from datasets import load_metric, load_dataset
stsb_metric = load_metric('glue', 'stsb')
stsb = load_dataset('glue', 'stsb')
```

6. Afterward, we must load both models:

```
import tensorflow_hub as hub
use_model = hub.load(
    "https://tfhub.dev/google/universal-sentence-
encoder/4")
from sentence_transformers import SentenceTransformer
distilroberta = SentenceTransformer(
                        'stsb-distilroberta-base-v2')
```

7. Both of these models provide embeddings for a given sentence. To compare the similarity between two sentences, we will use cosine similarity. The following function takes sentences as a batch and provides cosine similarity for each pair by utilizing USE:

```
import tensorflow as tf
import math
def use_sts_benchmark(batch):
  sts_encode1 = \
  tf.nn.l2_normalize(use_model(tf.
constant(batch['sentence1'])), axis=1)
  sts_encode2 = \
  tf.nn.l2_normalize(use_model(tf.
constant(batch['sentence2'])),   axis=1)
  cosine_similarities = \
              tf.reduce_sum(tf.multiply(sts_encode1,sts_
encode2),axis=1)
```

```
   clip_cosine_similarities = \
            tf.clip_by_value(cosine_similarities,-1.0, 1.0)
   scores = 1.0 - \
            tf.acos(clip_cosine_similarities) / math.pi
 return scores
```

8. With small modifications, the same function can be used for RoBERTa too. These small modifications are only for replacing the embedding function, which is different for TensorFlow Hub models and transformers. The following is the modified function:

```
def roberta_sts_benchmark(batch):
  sts_encode1 = \
  tf.nn.l2_normalize(distilroberta.
encode(batch['sentence1']), axis=1)
  sts_encode2 = \
    tf.nn.l2_normalize(distilroberta.
encode(batch['sentence2']), axis=1)
  cosine_similarities = \
           tf.reduce_sum(tf.multiply(sts_encode1, sts_
encode2),  axis=1)
  clip_cosine_similarities = tf.clip_by_value(cosine_
similarities, -1.0, 1.0)
  scores = 1.0  - tf.acos(clip_cosine_similarities) /
math.pi
 return scores
```

9. Applying these functions to the dataset will result in similarity scores for each of the models:

```
use_results = use_sts_benchmark(stsb['validation'])
distilroberta_results = roberta_sts_benchmark(
                                   stsb['validation'])
```

10. Using metrics on both results produces the Spearman and Pearson correlation values:

```
results = {
    "USE":stsb_metric.compute(
              predictions=use_results,
              references=references),
```

```
        "DistillRoberta":stsb_metric.compute(
                predictions=distilroberta_results,
                references=references)
    }
```

11. You can simply use pandas to see the results as a table in a comparative fashion:

```
import pandas as pd
pd.DataFrame(results)
```

The output is as follows:

	USE	DistillRoberta
pearson	0.810302	0.888461
spearmanr	0.808917	0.889246

Figure 7.2 – STSB validation results on DistilRoberta and USE

In this section, you learned about the important benchmarks of semantic textual similarity. Regardless of the model, you learned how to use any of these metrics to quantify model performance. In the next section, you will learn about the few-shot learning models.

Using BART for zero-shot learning

In the field of machine learning, zero-shot learning is referred to as models that can perform a task without explicitly being trained on it. In the case of NLP, it's assumed that there's a model that can predict the probability of some text being assigned to classes that are given to the model. However, the interesting part about this type of learning is that the model is not trained on these classes.

With the rise of many advanced language models that can perform transfer learning, zero-shot learning came to life. In the case of NLP, this kind of learning is performed by NLP models at test time, where the model sees samples belonging to new classes where no samples of them were seen before.

This kind of learning is usually used for classification tasks, where both the classes and the text are represented and the semantic similarity of both is compared. The represented form of these two is an embedding vector, while the similarity metric (such as cosine similarity or a pre-trained classifier such as a dense layer) outputs the probability of the sentence/text being classified as the class.

There are many methods and schemes we can use to train such models, but one of the earliest methods used crawled pages from the internet containing keyword tags in the meta part. For more information, read the following article and blog post at `https://amitness.com/2020/05/zero-shot-text-classification/`.

Instead of using such huge data, there are language models such as BART that use the **Multi-Genre Natural Language Inference** (**MNLI**) dataset to fine-tune and detect the relationship between two different sentences. Also, the HuggingFace model repository contains many models that have been implemented for zero-shot learning. They also provide a zero-shot learning pipeline for ease of use.

For example, BART from **Facebook AI Research** (**FAIR**) is being used in the following code to perform zero-shot text classification:

```
from transformers import pipeline
import pandas as pd
classifier = pipeline("zero-shot-classification",
                    model="facebook/bart-large-mnli")
sequence_to_classify = "one day I will see the world"
candidate_labels = ['travel',
                    'cooking',
                    'dancing',
                    'exploration']
result = classifier(sequence_to_classify, candidate_labels)
pd.DataFrame(result)
```

The results are as follows:

	sequence	labels	scores
0	one day I will see the world	travel	0.795756
1	one day I will see the world	exploration	0.199332
2	one day I will see the world	dancing	0.002621
3	one day I will see the world	cooking	0.002291

Figure 7.3 – Results of zero-shot learning using BART

As you can see, the travel and exploration labels have the highest probability, but the most probable one is travel.

However, sometimes, one sample can belong to more than one class (multilabel). HuggingFace provides a parameter called `multi_label` for this. The following example uses this parameter:

```
result = classifier(sequence_to_classify,
                     candidate_labels,
                     multi_label=True)
Pd.DataFrame(result)
```

Due to this, it is changed to the following:

	sequence	labels	scores
0	one day I will see the world	travel	0.994511
1	one day I will see the world	exploration	0.938389
2	one day I will see the world	dancing	0.005706
3	one day I will see the world	cooking	0.001819

Figure 7.4 – Results of zero-shot learning using BART (multi_label = True)

You can test the results even further and see how the model performs if very similar labels to the travel one are used. For example, you can see how it performs if `moving` and `going` are added to the label list.

There are other models that also leverage the semantic similarity between labels and the context to perform zero-shot classification. In the case of few-shot learning, some samples are given to the model, but these samples are not enough to train a model alone. Models can use these samples to perform tasks such as semantic text clustering, which will be explained shortly.

Now that you've learned how to use BART for zero-shot learning, you should learn how it works. BART is fine-tuned on **Natural Language Inference** (**NLI**) datasets such as MNLI. These datasets contain sentence pairs and three classes for each pair; that is, *Neutral*, *Entailment*, and *Contradiction*. Models that have been trained on these datasets can capture the semantics of two sentences and classify them by assigning a label in one-hot format. If you take out Neutral labels and only use Entailment and Contradiction as your output labels, if two sentences can come after each other, then it means these two are closely related to each other. In other words, you can change the first sentence to the label (`travel`, for example) and the second sentence to the content (`one day I will see the world`, for example). According to this, if these two can come after each other, this means that the label and the content are semantically related. The following code example shows how to directly use the BART model without the zero-shot classification pipeline according to the preceding descriptions:

```
from transformers \
     import AutoModelForSequenceClassification,\
     AutoTokenizer
nli_model = AutoModelForSequenceClassification\
               .from_pretrained(
                    "facebook/bart-large-mnli")
tokenizer = AutoTokenizer\
               .from_pretrained(
              "facebook/bart-large-mnli")
premise = "one day I will see the world"
label = "travel"
hypothesis = f'This example is {label}.'
x = tokenizer.encode(
    premise,
    hypothesis,
    return_tensors='pt',
    truncation_strategy='only_first')
logits = nli_model(x)[0]
entail_contradiction_logits = logits[:,[0,2]]
probs = entail_contradiction_logits.softmax(dim=1)
prob_label_is_true = probs[:,1]
print(prob_label_is_true)
```

The result is as follows:

```
tensor([0.9945], grad_fn=<SelectBackward>)
```

You can also call the first sentence the hypothesis and the sentence containing the label the premise. According to the result, the premise can entail the hypothesis. which means that the hypothesis is labeled as the premise.

So far, you've learned how to use zero-shot learning by utilizing NLI fine-tuned models. Next, you will learn how to perform few-/one-shot learning using semantic text clustering and semantic search.

Semantic similarity experiment with FLAIR

In this experiment, we will qualitatively evaluate the sentence representation models thanks to the `flair` library, which really simplifies obtaining the document embeddings for us.

We will perform experiments while taking on the following approaches:

- Document average pool embeddings
- RNN-based embeddings
- BERT embeddings
- SBERT embeddings

We need to install these libraries before we can start the experiments:

```
!pip install sentence-transformers
!pip install dataset
!pip install flair
```

For qualitative evaluation, we define a list of similar sentence pairs and a list of dissimilar sentence pairs (five pairs for each). What we expect from the embeddings models is that they should measure a high score and a low score, respectively.

The sentence pairs are extracted from the SBS Benchmark dataset, which we are already familiar with from the sentence-pair regression part of *Chapter 6, Fine-Tuning Language Models for Token Classification*. For similar pairs, two sentences are completely equivalent, and they share the same meaning.

The pairs with a similarity score of around 5 in the STSB dataset are randomly taken, as follows:

```
import pandas as pd
similar=[
 ("A black dog walking beside a pool.",
 "A black dog is walking along the side of a pool."),
 ("A blonde woman looks for medical supplies for work in a
suitcase. ",
 " The blond woman is searching for medical supplies in a
suitcase."),
 ("A doubly decker red bus driving down the road.",
 "A red double decker bus driving down a street."),
 ("There is a black dog jumping into a swimming pool.",
 "A black dog is leaping into a swimming pool."),
 ("The man used a sword to slice a plastic bottle.",
 "A man sliced a plastic bottle with a sword.")]
pd.DataFrame(similar, columns=["sen1", "sen2"])
```

The output is as follows:

	sen1	sen2
0	A black dog walking beside a pool.	A black dog is walking along the side of a pool.
1	A blonde woman looks for medical supplies for ...	The blond woman is searching for medical supp...
2	A doubly decker red bus driving down the road.	A red double decker bus driving down a street.
3	There is a black dog jumping into a swimming p...	A black dog is leaping into a swimming pool.
4	The man used a sword to slice a plastic bottle.\t	A man sliced a plastic bottle with a sword.

Figure 7.5 – Similar pair list

Here is the list of dissimilar sentences whose similarity scores are around 0, taken from the STS-B dataset:

```
import pandas as pd
dissimilar= [
 ("A little girl and boy are reading books. ",
```

```
    "An older child is playing with a doll while gazing out the
window."),
    ("Two horses standing in a field with trees in the
background.",
    "A black and white bird on a body of water with grass in the
background."),
    ("Two people are walking by the ocean.",
    "Two men in fleeces and hats looking at the camera."),
    ("A cat is pouncing on a trampoline.",
    "A man is slicing a tomato."),
    ("A woman is riding on a horse.",
    "A man is turning over tables in anger.")]
    pd.DataFrame(dissimilar, columns=["sen1", "sen2"])
```

The output is as follows:

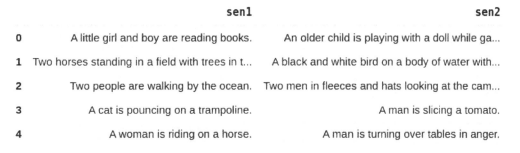

	sen1	sen2
0	A little girl and boy are reading books.	An older child is playing with a doll while ga...
1	Two horses standing in a field with trees in t...	A black and white bird on a body of water with...
2	Two people are walking by the ocean.	Two men in fleeces and hats looking at the cam...
3	A cat is pouncing on a trampoline.	A man is slicing a tomato.
4	A woman is riding on a horse.	A man is turning over tables in anger.

Figure 7.6 – Dissimilar pair list

Now, let's prepare the necessary functions to evaluate the embeddings models. The following sim() function computes the cosine similarity between two sentences; that is, s1, s2:

```
import torch, numpy as np
def sim(s1,s2):
    s1=s1.embedding.unsqueeze(0)
    s2=s2.embedding.unsqueeze(0)
    sim=torch.cosine_similarity(s1,s2).item()
    return np.round(sim,2)
```

The document embeddings models that were used in this experiment are all pre-trained models. We will pass the document embeddings model object and sentence pair list (similar or dissimilar) to the following `evaluate()` function, where, once the model encodes the sentence embeddings, it will compute the similarity score for each pair in the list, along with the list average. The definition of the function is as follows:

```
from flair.data import Sentence
def evaluate(embeddings, myPairList):
    scores=[]
    for s1, s2 in myPairList:
        s1,s2=Sentence(s1), Sentence(s2)
        embeddings.embed(s1)
        embeddings.embed(s2)
        score=sim(s1,s2)
        scores.append(score)
    return scores, np.round(np.mean(scores),2)
```

Now, it is time to evaluate sentence embedding models. We will start with the average pooling method!

Average word embeddings

Average word embeddings (or **document pooling**) apply the mean pooling operation to all the words in a sentence, where the average of all the word embeddings is considered to be sentence embedding. The following execution instantiates a document pool embedding based on GloVe vectors. Note that although we will use only GloVe vectors here, the flair API allows us to use multiple word embeddings. Here is the code definition:

```
from flair.data import Sentence
from flair.embeddings\
        import WordEmbeddings, DocumentPoolEmbeddings
glove_embedding = WordEmbeddings('glove')
glove_pool_embeddings = DocumentPoolEmbeddings(
                                        [glove_embedding]
                                        )
```

Let's evaluate the GloVe pool model on similar pairs, as follows:

```
>>> evaluate(glove_pool_embeddings, similar)
([0.97, 0.99, 0.97, 0.99, 0.98], 0.98)
```

The results seem to be good since those resulting values are very high, which is what we expect. However, the model produces high scores such as 0.94 on average for the dissimilar list as well. Our expectation would be less than 0.4. We'll talk about why we got this later in this chapter. Here is the execution:

```
>>> evaluate(glove_pool_embeddings, dissimilar)
([0.94, 0.97, 0.94, 0.92, 0.93], 0.94)
```

Next, let's evaluate some RNN embeddings on the same problem.

RNN-based document embeddings

Let's instantiate a GRU model based on GloVe embeddings, where the default model of DocumentRNNEmbeddings is a GRU:

```
from flair.embeddings \
        import WordEmbeddings, DocumentRNNEmbeddings
gru_embeddings = DocumentRNNEmbeddings([glove_embedding])
```

Run the evaluation method:

```
>>> evaluate(gru_embeddings, similar)
([0.99, 1.0, 0.94, 1.0, 0.92], 0.97)
>>> evaluate(gru_embeddings, dissimilar)
([0.86, 1.0, 0.91, 0.85, 0.9], 0.9)
```

Likewise, we get a high score for the dissimilar list. This is not what we want from sentence embeddings.

Transformer-based BERT embeddings

The following execution instantiates a bert-base-uncased model that pools the final layer:

```
from flair.embeddings import TransformerDocumentEmbeddings
from flair.data import Sentence
bert_embeddings = TransformerDocumentEmbeddings(
                                'bert-base-uncased')
```

Run the evaluation, as follows:

```
>>> evaluate(bert_embeddings, similar)
([0.85, 0.9, 0.96, 0.91, 0.89], 0.9)
>>> evaluate(bert_embeddings, dissimilar)
([0.93, 0.94, 0.86, 0.93, 0.92], 0.92)
```

This is worse! The score of the dissimilar list is higher than that of the similar list.

Sentence-BERT embeddings

Now, let's apply Sentence-BERT to the problem of distinguishing similar pairs from dissimilar ones, as follows:

1. First of all, a warning: we need to ensure that the sentence-transformers package has already been installed:

    ```
    !pip install sentence-transformers
    ```

2. As we mentioned previously, Sentence-BERT provides a variety of pre-trained models. We will pick the bert-base-nli-mean-tokens model for evaluation. Here is the code:

    ```
    from flair.data import Sentence
    from flair.embeddings \
            import SentenceTransformerDocumentEmbeddings
    sbert_embeddings = SentenceTransformerDocumentEmbeddings(
                      'bert-base-nli-mean-tokens')
    ```

3. Let's evaluate the model:

    ```
    >>> evaluate(sbert_embeddings, similar)
    ([0.98, 0.95, 0.96, 0.99, 0.98], 0.97)
    >>> evaluate(sbert_embeddings, dissimilar)
    ([0.48, 0.41, 0.19, -0.05, 0.0], 0.21)
    ```

Well done! The SBERT model produced better results. The model produced a low similarity score for the dissimilar list, which is what we expect.

4. Now, we will do a harder test, where we pass contradicting sentences to the models. We will define some tricky sentence pairs, as follows:

```
>>> tricky_pairs=[
("An elephant is bigger than a lion",
"A lion is bigger than an elephant") ,
("the cat sat on the mat",
"the mat sat on the cat")]
>>> evaluate(glove_pool_embeddings, tricky_pairs)
([1.0, 1.0], 1.0)
>>> evaluate(gru_embeddings, tricky_pairs)
([0.87, 0.65], 0.76)
>>> evaluate(bert_embeddings, tricky_pairs)
([1.0, 0.98], 0.99)
>>> evaluate(sbert_embeddings, tricky_pairs)
([0.93, 0.97], 0.95)
```

Interesting! The scores are very high since the sentence similarity model works similar to topic detection and measures content similarity. When we look at the sentences, they share the same content, even though they contradict each other. The content is about lion and elephant or cat and mat. Therefore, the models produce a high similarity score. Since the GloVe embedding method pools the average of the words without caring about word order, it measures two sentences as being the same. On the other hand, the GRU model produced lower values as it cares about word order. Surprisingly, even the SBERT model does not produce efficient scores. This may be due to the content similarity-based supervision that's used in the SBERT model.

5. To correctly detect the semantics of two sentence pairs with three classes – that is, Neutral, Contradiction, and Entailment – we must use a fine-tuned model on MNLI. The following code block shows an example of using XLM-Roberta, fine-tuned on XNLI with the same examples:

```
from transformers \
Import AutoModelForSequenceClassification, AutoTokenizer
nli_model = AutoModelForSequenceClassification\
                .from_pretrained(
                    'joeddav/xlm-roberta-large-xnli')
tokenizer = AutoTokenizer\
```

```
                 .from_pretrained(
                      'joeddav/xlm-roberta-large-xnli')
import numpy as np
for permise, hypothesis in tricky_pairs:
  x = tokenizer.encode(premise,
                       hypothesis,
                       return_tensors='pt',
                       truncation_strategy='only_first')
    logits = nli_model(x)[0]
    print(f"Permise: {permise}")
    print(f"Hypothesis: {hypothesis}")
    print("Top Class:")
   print(nli_model.config.id2label[np.argmax(
                      logits[0].detach().numpy()). ])
    print("Full softmax scores:")
    for i in range(3):
      print(nli_model.config.id2label[i],
            logits.softmax(dim=1)[0][i].detach().numpy())
    print("="*20)
```

6. The output will show the correct labels for each:

```
Permise: An elephant is bigger than a lion
Hypothesis: A lion is bigger than an elephant
Top Class:
contradiction
Full softmax scores:
contradiction 0.7731286
neutral 0.2203285
entailment 0.0065428796

====================
Permise: the cat sat on the mat
Hypothesis: the mat sat on the cat
Top Class:
entailment
Full softmax scores:
contradiction 0.49365467
```

```
neutral 0.007260764
entailment 0.49908453

=====================
```

In some problems, In some problems, NLI is a higher priority than semantic textual because it is intended to find the contradiction or entailment rather than the raw similarity score. For the next sample, use two sentences for entailment and contradiction at the same time. This is a bit subjective, but to the model, the second sentence pair seems to be a very close call between entailment and contradiction.

Text clustering with Sentence-BERT

For clustering algorithms, we will need a model that's suitable for textual similarity. Let's use the `paraphrase-distilroberta-base-v1` model here for a change. We will start by loading the Amazon Polarity dataset for our clustering experiment. This dataset includes Amazon web page reviews spanning a period of 18 years up to March 2013. The original dataset includes over 35 million reviews. These reviews include product information, user information, user ratings, and user reviews. Let's get started:

1. First, randomly select 10K reviews by shuffling, as follows:

    ```
    import pandas as pd, numpy as np
    import torch, os, scipy
    from datasets import load_dataset
    dataset = load_dataset("amazon_polarity",split="train")
    corpus=dataset.shuffle(seed=42)[:10000]['content']
    ```

2. The corpus is now ready for clustering. The following code instantiates a sentence-transformer object using the pre-trained `paraphrase-distilroberta-base-v1` model:

    ```
    from sentence_transformers import SentenceTransformer
    model_path="paraphrase-distilroberta-base-v1"
    model = SentenceTransformer(model_path)
    ```

3. The entire corpus is encoded with the following execution, where the model maps a list of sentences to a list of embedding vectors:

    ```
    >>> corpus_embeddings = model.encode(corpus)
    >>> corpus_embeddings.shape
    (10000, 768)
    ```

4. Here, the vector size is `768`, which is the default embedding size of the BERT-base model. From now on, we will proceed with traditional clustering methods. We will choose *Kmeans* here since it is a fast and widely used clustering algorithm. We just need to set the cluster number (*K*) to 5. Actually, this number may not be optimal. There are many techniques that can determine the optimal number of clusters, such as the Elbow or Silhouette method. However, let's leave these issues aside. Here is the execution:

```
>>> from sklearn.cluster import KMeans
>>> K=5
>>> kmeans = KMeans(
            n_clusters=5,
            random_state=0).fit(corpus_embeddings)
>>> cls_dist=pd.Series(kmeans.labels_).value_counts()
>>> cls_dist
3 2772
4 2089
0 1911
2 1883
1 1345
```

Here, we have obtained five clusters of reviews. As we can see from the output, we have fairly distributed clusters. Another issue with clustering is that we need to understand what these clusters mean. As a suggestion, we can apply topic analysis to each cluster or check cluster-based TF-IDF to understand the content. Now, let's look at another way to do this based on the cluster centers. The Kmeans algorithm computes cluster centers, called centroids, that are kept in the `kmeans.cluster_centers_` attribute. The centroids are simply the average of the vectors in each cluster. Therefore, they are all imaginary points, not the existing data points. Let's assume that the sentences closest to the centroid will be the most representative example for the corresponding cluster.

5. Let's try to find only one real sentence embedding, closest to each centroid point. If you like, you can capture more than one sentence. Here is the code:

```
distances = \
scipy.spatial.distance.cdist(kmeans.cluster_centers_,
corpus_embeddings)
centers={}
print("Cluster", "Size", "Center-idx",
```

```
                          "Center-Example", sep="\t\t")
     for i,d in enumerate(distances):
          ind = np.argsort(d, axis=0)[0]
          centers[i]=ind
          print(i,cls_dist[i], ind, corpus[ind] ,sep="\t\t")
```

The output is as follows:

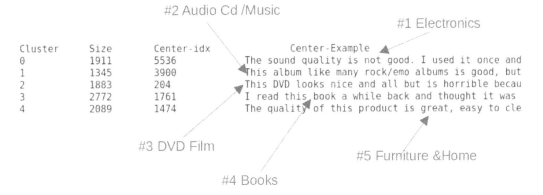

Figure 7.7 – Centroids of the cluster

From these representative sentences, we can reason about the clusters. It seems to be that Kmeans clusters the reviews into five distinct categories: *Electronics, Audio Cd/ Music, DVD Film, Books,* and *Furniture & Home.* Now, let's visualize both sentence points and cluster centroids in a 2D space. We will use the **Uniform Manifold Approximation and Projection (UMAP)** library to reduce dimensionality. Other widely used dimensionality reduction techniques in NLP that you can use include t-SNE and PCA (see *Chapter 1, From Bag-of-Words to the Transformers*).

6. We need to install the umap library, as follows:

    ```
    !pip install umap-learn
    ```

7. The following execution reduces all the embeddings and maps them into a 2D space:

    ```
    import matplotlib.pyplot as plt
    import umap
    X = umap.UMAP(
             n_components=2,
             min_dist=0.0).fit_transform(corpus_embeddings)
    labels= kmeans.labels_fig, ax = plt.subplots(figsize=(12,
    ```

```
8))
plt.scatter(X[:,0], X[:,1], c=labels, s=1, cmap='Paired')
for c in centers:
    plt.text(X[centers[c],0], X[centers[c], 1],"CLS-"+
str(c), fontsize=18)
    plt.colorbar()
```

The output is as follows:

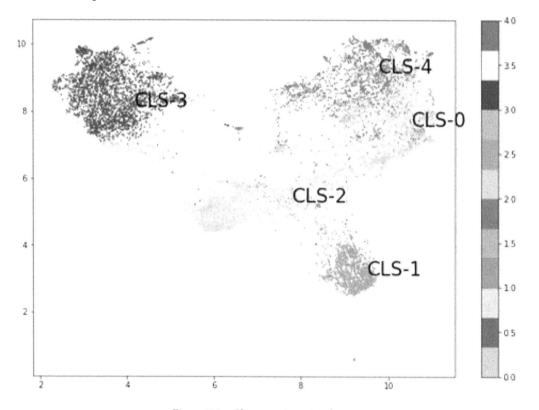

Figure 7.8 – Cluster points visualization

In the preceding output, the points have been colored according to their cluster membership and centroids. It looks like we have picked the right number of clusters.

To capture the topics and interpret the clusters, we simply located the sentences (one single sentence for each cluster) close to the centroids of the clusters. Now, let's look at a more accurate way of capturing the topic with topic modeling.

Topic modeling with BERTopic

You may be familiar with many unsupervised topic modeling techniques that are used to extract topics from documents; **Latent-Dirichlet Allocation** (**LDA**) topic modeling and **Non-Negative Matrix Factorization** (**NMF**) are well-applied traditional techniques in the literature. BERTopic and Top2Vec are two important transformer-based topic modeling projects. In this section, we will apply the BERTopic model to our Amazon corpus. It leverages BERT embeddings and the class-based TF-IDF method to get easily interpretable topics.

First, the BERTopic model starts by encoding the sentences with sentence transformers or any sentence embedding model, which is followed by the clustering step. The clustering step has two phases: the embedding's dimensionality is reduced by **UMAP** and then the reduced vectors are clustered by **Hierarchical Density-Based Spatial Clustering of Applications with Noise** (**HDBSCAN**), which yields groups of similar documents. At the final stage, the topics are captured by cluster-wise TF-IDF, where the model extracts the most important words per cluster rather than per document and obtains descriptions of the topics for each cluster. Let's get started:

1. First, let's install the necessary library, as follows:

    ```
    !pip install bertopic
    ```

    > **Important note**
    > You may need to restart the runtime since this installation will update some packages that have already been loaded. So, from the Jupyter notebook, go to **Runtime | Restart Runtime**.

2. If you want to use your own embedding model, you need to instantiate and pass it through the BERTopic model. We will instantiate a Sentence Transformer model and pass it to the constructor of BERTopic, as follows:

    ```
    from bertopic import BERTopic
    sentence_model = SentenceTransformer(
                    "paraphrase-distilroberta-base-v1")
    topic_model = BERTopic(embedding_model=sentence_model)
    topics, _ = topic_model.fit_transform(corpus)
    topic_model.get_topic_info()[:6]
    ```

The output is as follows:

	Topic	Count	Name
0	4	3086	4_book_read_books_who
1	-1	1818	-1_product_my_use_have
2	7	1499	7_movie_film_dvd_watch
3	5	1327	5_album_cd_songs_music
4	24	274	24_toy_daughter_we_loves
5	2	235	2_game_games_play_graphics

Figure 7.9 – BERTopic results

Please note that different BERTopic runs with the same parameters can yield different results since the UMAP model is stochastic. Now, let's see the word distribution of topic five, as follows:

```
topic_model.get_topic(5)
```

The output is as follows:

```
[('album', 0.021777776441862785),
 ('cd', 0.0216003728561258),
 ('songs', 0.015716979809362878),
 ('music', 0.015336261401310738),
 ('song', 0.012883049138010031),
 ('band', 0.0087909916825825062),
 ('great', 0.006907063839145953),
 ('good', 0.006594220889305517),
 ('he', 0.006428544176459775),
 ('albums', 0.006402900278216675)]
```

Figure 7.10 – The fifth topic words of the topic model

The topic words are those words whose vectors are close to the topic vector in the semantic space. In this experiment, we did not cluster the corpus; instead, we applied the technique to the entire corpus. In our previous example, we analyzed the clusters with the closest sentence. Now, we can find the topics by applying the topic model separately to each cluster. This is pretty straightforward, and you can run it yourself.

Please see the Top2Vec project for more details and interesting topic modeling applications at https://github.com/ddangelov/Top2Vec.

Semantic search with Sentence-BERT

We may already be familiar with keyword-based search (Boolean model), where, for a given keyword or pattern, we can retrieve the results that match the pattern. Alternatively, we can use regular expressions, where we can define advanced patterns such as the lexico-syntactic pattern. These traditional approaches cannot handle synonym (for example, *car* is the same as *automobile*) or word sense problems (for example, *bank* as the side of a river or *bank* as a financial institute). While the first synonym case causes low recall due to missing out the documents that shouldn't be missed, the second causes low precision due to catching the documents not to be caught. Vector-based or semantic search approaches can overcome these drawbacks by building a dense numerical representation of both queries and documents.

Let's set up a case study for **Frequently Asked Questions** (**FAQs**) that are idle on websites. We will exploit FAQ resources within a semantic search problem. FAQs contain frequently asked questions. We will be using the FAQ from the **World Wide Fund for Nature** (**WWF**), a nature non-governmental organization (`https://www.wwf.org.uk/`).

Given these descriptions, it is easy to understand that performing a semantic search using semantic models is very similar to a one-shot learning problem, where we just have a single shot of the class (a single sample), and we want to reorder the rest of the data (sentences) according to it. You can redefine the problem as searching for samples that are semantically close to the given sample, or a binary classification according to the sample. Your model can provide a similarity metric, and the results for all the other samples will be reordered using this metric. The final ordered list is the search result, which is reordered according to semantic representation and the similarity metric.

WWF has 18 questions and answers on their web page. We defined them as a Python list object called `wf_faq` for this experiment:

- I haven't received my adoption pack. What should I do?
- How quickly will I receive my adoption pack?
- How can I renew my adoption?
- How do I change my address or other contact details?
- Can I adopt an animal if I don't live in the UK?
- If I adopt an animal, will I be the only person who adopts that animal?
- My pack doesn't contain a certificate?
- My adoption is a gift but won't arrive on time. What can I do?
- Can I pay for an adoption with a one-off payment?
- Can I change the delivery address for my adoption pack after I've placed my order?

- How long will my adoption last for?

- How often will I receive updates about my adopted animal?

- What animals do you have for adoption?

- How can I find out more information about my adopted animal?

- How is my adoption money spent?

- What is your refund policy?

- An error has been made with my Direct Debit payment; can I receive a refund?

- How do I change how you contact me?

Users are free to ask any question they want. We need to evaluate which question in the FAQ is the most similar to the user's question, which is the objective of the `quora-distilbert-base` model. There are two options in the SBERT hub – one is for English and another for multilingual, as follows:

- `quora-distilbert-base`: This is fine-tuned for Quora Duplicate Questions detection retrieval.

- `quora-distilbert-multilingual`: This is a multilingual version of `quora-distilbert-base`. It's fine-tuned with parallel data for 50+ languages.

Let's build a semantic search model by following these steps:

1. The following is the SBERT model's instantiation:

```
from sentence_transformers import SentenceTransformer
model = SentenceTransformer('quora-distilbert-base')
```

2. Let's encode the FAQ, as follows:

```
faq_embeddings = model.encode(wwf_faq)
```

3. Let's prepare five questions so that they are similar to the first five questions in the FAQ, respectively; that is, our first test question should be similar to the first question in the FAQ, the second question should be similar to the second question, and so on, so that we can easily follow the results. Let's define the questions in the `test_questions` list object and encode it, as follows:

```
test_questions=["What should be done, if the adoption
pack did not reach to me?",
" How fast is my adoption pack delivered to me?",
```

```
"What should I do to renew my adoption?",
"What should be done to change address and contact
details ?",
"I live outside of the UK, Can I still adopt an animal?"]
test_q_emb= model.encode(test_questions)
```

4. The following code measures the similarity between each test question and each
    question in the FAQ and then ranks them:

```
from scipy.spatial.distance import cdist
for q, qe in zip(test_questions, test_q_emb):
    distances = cdist([qe], faq_embeddings, "cosine")[0]
    ind = np.argsort(distances, axis=0)[:3]
    print("\n Test Question: \n "+q)
    for i,(dis,text) in enumerate(
                                zip(
                                distances[ind],
                                [wwf_faq[i] for i in ind])):
        print(dis,ind[i],text, sep="\t")
```

The output is as follows:

```
Test Question:
What should be done, if the adoption pack did not reach to me?
0.1494580342947357      0    I haven't received my adoption pack. What should I do?
0.24940214249978787     7    My adoption is a gift but won't arrive on time. What can I do?
0.3669761157176866      1    How quickly will I receive my adoption pack?

Test Question:
 How fast is my adoption pack delivered to me?
0.16582390267585112     1    How quickly will I receive my adoption pack?
0.3470478678903325      0    I haven't received my adoption pack. What should I do?
0.3511114386193057      7    My adoption is a gift but won't arrive on time. What can I do?

Test Question:
What should I do to renew my adoption?
0.04168242777718267     2    How can I renew my adoption?
0.2993018812386016      12   What animals do you have for adoption?
0.3014071168242859      0    I haven't received my adoption pack. What should I do?

Test Question:
What should be done to change adress and contact details ?
0.276601898726506       3    How do I change my address or other contact details?
0.352868128705782       17   How do I change how you contact me?
0.4393553216276348      2    How can I renew my adoption?

Test Question:
I live outside of the UK, Can I still adopt an animal?
0.16945626472973518     4    Can I adopt an animal if I don't live in the UK?
0.200544029334076       12   What animals do you have for adoption?
0.28782233378715627     13   How can I nd out more information about my adopted animal?
```

Figure 7.11 – Question-question similarity

Here, we can see indexes 0, 1, 2, 3, and 4 in order, which means the model successfully found the similar questions as expected.

5. For the deployment, we can design the following `getBest()` function, which takes a question and returns K most similar questions in the FAQ:

```
def get_best(query, K=5):
    query_emb = model.encode([query])
    distances = cdist(query_emb,faq_embeddings,"cosine")
[0]
    ind = np.argsort(distances, axis=0)
    print("\n"+query)
    for c,i in list(zip(distances[ind], ind))[:K]:
        print(c,wwf_faq[i], sep="\t")
```

6. Let's ask a question:

```
get_best("How do I change my contact info?",3)
```

The output is as follows:

```
How do I change my contact info?
0.05676792449319612    How do I change my address or other contact details?
0.185665422885958      How do I change how you contact me?
0.32408327251343816    How can I renew my adoption?
```

Figure 7.12 – Similar question similarity results

7. What if a question that's used as input is not similar to one from the FAQ? Here is such a question:

```
get_best("How do I get my plane ticket \
    if I bought it online?")
```

The output is as follows:

```
How do I get my plane ticket if I bought it online?
0.35947505490536136    How do I change how you contact me?
0.3680785568009698     How do I change my address or other contact details?
0.4306634329555338     My adoption is a gift but won't arrive on time. What can I do?
```

Figure 7.13 – Dissimilar question similarity results

The best dissimilarity score is 0.35. So, we need to define a threshold such as 0.3 so that the model ignores such questions that are higher than that threshold and says `no similar answer found`.

Other than question-question symmetric search similarity, we can also utilize SBERT's question-answer asymmetric search models, such as `msmarco-distilbert-base-v3`, which is trained on a dataset of around 500K Bing search queries. It is known as MSMARCO Passage Ranking. This model helps us measure how related the question and context are and checks whether the answer to the question is in the passage.

Summary

In this chapter, we learned about text representation methods. We learned how it is possible to perform tasks such as zero-/few-/one-shot learning using different and diverse semantic models. We also learned about NLI and its importance in capturing semantics of text. Moreover, we looked at some useful use cases such as semantic search, semantic clustering, and topic modeling using Transformer-based semantic models. We learned how to visualize the clustering results and understood the importance of centroids in such problems.

In the next chapter, you will learn about efficient Transformer models. You will learn about distillation, pruning, and quantizing Transformer-based models. You will also learn about different and efficient Transformer architectures that make improvements to computational and memory efficiency, as well as how to use them in NLP problems.

Further reading

Please refer to the following works/papers for more information about the topics that were covered in this chapter:

- Lewis, M., Liu, Y., Goyal, N., Ghazvininejad, M., Mohamed, A., Levy, O., ... & Zettlemoyer, L. (2019). *Bart: Denoising sequence-to-sequence pre-training for natural language generation, translation, and comprehension.* arXiv preprint arXiv:1910.13461.

- Pushp, P. K., & Srivastava, M. M. (2017). *Train once, test anywhere: Zero-shot learning for text classification.* arXiv preprint arXiv:1712.05972.

- Reimers, N., & Gurevych, I. (2019). *Sentence-bert: Sentence embeddings using siamese bert-networks.* arXiv preprint arXiv:1908.10084.

- Liu, Y., Ott, M., Goyal, N., Du, J., Joshi, M., Chen, D., ... & Stoyanov, V. (2019). *Roberta: A robustly optimized bert pretraining approach.* arXiv preprint arXiv:1907.11692.

- Williams, A., Nangia, N., & Bowman, S. R. (2017). *A broad-coverage challenge corpus for sentence understanding through inference.* arXiv preprint arXiv:1704.05426.

- Cer, D., Yang, Y., Kong, S. Y., Hua, N., Limtiaco, N., John, R. S., ... & Kurzweil, R. (2018). *Universal sentence encoder.* arXiv preprint arXiv:1803.11175.

- Yang, Y., Cer, D., Ahmad, A., Guo, M., Law, J., Constant, N., ... & Kurzweil, R. (2019). *Multilingual universal sentence encoder for semantic retrieval.* arXiv preprint arXiv:1907.04307.

- Humeau, S., Shuster, K., Lachaux, M. A., & Weston, J. (2019). *Poly-encoders: Transformer architectures and pre-training strategies for fast and accurate multi-sentence scoring.* arXiv preprint arXiv:1905.01969.

Section 3: Advanced Topics

On completion of this section, you will have gained experience in how to train efficient models for challenging problems such as long-context NLP tasks under limited computational capacity, and how to work with multilingual and cross-lingual language modeling. You will learn about the tools needed to monitor the inner parts of the models for explainability and interpretability and to track your model-training performance. You will also be able to serve models in a real production environment.

This section comprises the following chapters:

- *Chapter 8, Working with Efficient Transformers*
- *Chapter 9, Cross-Lingual and Multilingual Language Modeling*
- *Chapter 10, Serving Transformer Models*
- *Chapter 11, Attention Visualization and Experiment Tracking*

8
Working with Efficient Transformers

So far, you have learned how to design a **Natural Language Processing** (NLP) architecture to achieve successful task performance with transformers. In this chapter, you will learn how to make efficient models out of trained models using distillation, pruning, and quantization. Second, you will also gain knowledge about efficient sparse transformers such as Linformer, BigBird, Performer, and so on. You will see how they perform on various benchmarks, such as memory versus sequence length and speed versus sequence length. You will also see the practical use of model size reduction.

The importance of this chapter came to light as it is getting difficult to run large neural models under limited computational capacity. It is important to have a lighter general-purpose language model such as DistilBERT. This model can then be fine-tuned with good performance, like its non-distilled counterparts. Transformers-based architectures face complexity bottlenecks due to the quadratic complexity of the attention dot product in the transformers, especially for long-context NLP tasks. Character-based language models, speech processing, and long documents are among the long-context problems. In recent years, we have seen much progress in making self-attention more efficient, such as Reformer, Performer, and BigBird, as a solution to complexity.

In short, in this chapter, you will learn about the following topics:

- Introduction to efficient, light, and fast transformers
- Implementation for model size reduction
- Working with efficient self-attention

Technical requirements

We will be using the Jupyter Notebook to run our coding exercises, which require Python 3.6+, and the following packages need to be installed:

- TensorFlow
- PyTorch
- Transformers >=4.00
- Datasets
- sentence-transformers
- py3nvml

All notebooks with coding exercises are available at the following GitHub link:

`https://github.com/PacktPublishing/Mastering-Transformers/tree/main/CH08`

Check out the following link to see Code in Action Video:

`https://bit.ly/3y5j9oZ`

Introduction to efficient, light, and fast transformers

Transformer-based models have distinctly achieved state-of-the-art results in many NLP problems at the cost of quadratic memory and computational complexity. We can highlight the issues regarding complexity as follows:

- The models are not able to efficiently process long sequences due to their self-attention mechanism, which scales quadratically with the sequence length.
- An experimental setup using a typical GPU with 16 GB can handle the sentences of 512 tokens for training and inference. However, longer entries can cause problems.

- The NLP models keep growing from the 110 million parameters of BERT-base to the 17 billion parameters of Turing-NLG and to the 175 billion parameters of GPT-3. This notion raises concerns about computational and memory complexity.

- We also need to care about costs, production, reproducibility, and sustainability. Hence, we need faster and lighter transformers, especially on edge devices.

Several approaches have been proposed to reduce computational complexity and memory footprint. Some of these approaches focus on changing the architecture and some do not alter the original architecture but instead make improvements to the trained model or to the training phase. We will divide them into two groups, model size reduction and efficient self-attention.

Model size reduction can be accomplished using three different approaches:

- Knowledge distillation

- Pruning

- Quantization

Each of these three has its own way of reducing the size model, which we will describe in short in the *Implementation for model size reduction* section.

In knowledge distillation, a smaller transformer (student) can transfer the knowledge of a big model (teacher). We train the student model so that it can mimic the teacher's behavior or produce the same output for the same input. The distilled model may underperform the teacher. There is a trade-off between compression, speed, and performance.

Pruning is a model compression technique in machine learning that is used to reduce the size of the model by removing a section of the model that contributes little to producing results. The most typical example is decision tree pruning, which helps to reduce the model complexity and increase the generalization capacity of the model. Quantization changes model weight types from higher resolutions to lower resolutions. For example, we use a typical floating-point number (`float64`) consuming 64 bits of memory for each weight. Instead, we can use `int8` in quantization, which consumes 8 bits for each weight, and naturally has less accuracy in presenting numbers.

Self-attention heads are not optimized for long sequences. In order to solve this issue, many different approaches have been proposed. The most efficient approach is **Self-Attention Sparsification**, which we will discuss soon. The other most widely used approach is **Memory Efficient Backpropagation**. This approach balances a trade-off between the caching of intermediate results and re-computing. Intermediate activations computed during forward propagation are needed to compute gradients during backward propagation. Gradient checkpoints can reduce a substantial amount of memory footprint and computation. Another approach is **Pipeline Parallelism Algorithms**. Mini-batches are split into micro-batches and the parallelism pipeline takes advantage of using the waiting time during the forward and backward operations while transferring the batches to deep learning accelerators such as **Graphics Processing Unit (GPU)** or **Tensor Processing Unit (TPU)**.

Parameter Sharing can be counted as one of the first approaches towards efficient deep learning. The most typical example is RNN, as depicted in *Chapter 1, From Bag-of-Words to the Transformers*, where the units of unfolded representation use the shared parameters. Hence, the number of trainable parameters is not affected by the input size. Some shared parameters which are also called weight tying or weight replication, spread the network so that the number of trainable parameters is reduced. For instance, Linformer shares projection matrices across heads and layers. Reformer shares the query and key at the cost of performance loss.

Now let's try to understand these issues with corresponding practical examples.

Implementation for model size reduction

Even though the transformer-based models achieve state-of-the-art results in many aspects of NLP, they usually share the very same problem: they are big models and are not fast enough to be used. In business cases where it is necessary to embed them inside a mobile application or in a web interface, it seems to be impossible if you try to use the original models.

In order to improve the speed and size of these models, some techniques are proposed, which are listed here:

- Distillation (also known as knowledge distillation)
- Pruning
- Quantization

For each of these techniques, we provide a separate subsection to address the technical and theoretical insights.

Working with DistilBERT for knowledge distillation

The process of transferring knowledge from a bigger model to a smaller one is called **knowledge distillation**. In other words, there is a teacher model and a student model; the teacher is typically a bigger and stronger model while the student is smaller and weaker.

This technique is used in various problems, from vision to acoustic models and NLP. A typical implementation of this technique is shown in *Figure 8.1*:

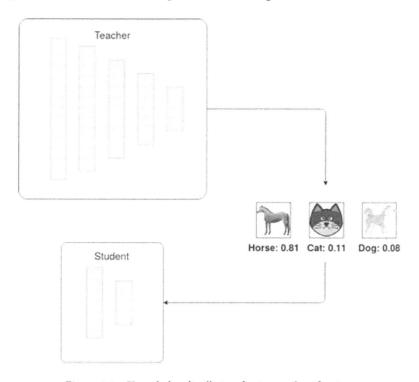

Figure 8.1 – Knowledge distillation for image classification

DistilBERT is one of the most important models in this field and has got the attention of researchers and even the businesses. This model, which tries to mimic the behavior of BERT-Base, has 50% fewer parameters and achieves 95% of the teacher model's performance.

Some details are given as follows:

- DistilBert is 1.7x compressed and 1.6x faster with 97% relative performance (compared to original BERT).

- Mini-BERT is 6x compressed, 3x faster, and has 98% relative performance.

- TinyBERT is 7.5x compressed, has 9.4x speed, and 97% relative performance.

The distillation training step that is used to train the model is very simple using PyTorch (original description and code available at `https://medium.com/huggingface/distilbert-8cf3380435b5`):

```python
import torch
import torch.nn as nn
import torch.nn.functional as F
from torch.optim import Optimizer
KD_loss = nn.KLDivLoss(reduction='batchmean')
def kd_step(teacher: nn.Module,
            student: nn.Module,
            temperature: float,
            inputs: torch.tensor,
            optimizer: Optimizer):
    teacher.eval()
    student.train()
    with torch.no_grad():
        logits_t = teacher(inputs=inputs)
    logits_s = student(inputs=inputs)
    loss = KD_loss(input=F.log_softmax(
                        logits_s/temperature,
                        dim=-1),
                    target=F.softmax(
                        logits_t/temperature,
                        dim=-1))
    loss.backward()
    optimizer.step()
    optimizer.zero_grad()
```

This model-supervised training provides us with a smaller model that is very similar to the base model in behavior. However, the loss function used here is **Kullback-Leibler** loss to ensure that the student model mimics the good and bad aspects of the teacher model with no modification of the decision on the last softmax logits. This loss function shows how different two distributions are from each other; a greater difference means a higher loss value. The reason for using this loss function is to make the student model try to completely mimic the behavior of the teacher. The GLUE macro scores for BERT and DistilBERT are just 2.8% different.

Pruning transformers

Pruning includes the process of setting weights at each layer to zero based on a pre-specified criterion. For example, a simple pruning algorithm could take the weights of each layer and set those that are below a threshold. This method eliminates weights that are very low in value and do not affect the results too much.

Likewise, we prune some redundant parts of the transformer network. The pruned networks are more likely to generalize better than the original one. We have seen a successful pruning operation because the pruning process probably keeps the true underlying explanatory factors and discards the redundant subnetwork. But we need to still train a large network. The reasonable strategy is that we train a neural network as large as possible. Then, the less salient weights or units whose removals have a small effect on the model performance are discarded.

There are two approaches:

- **Unstructured pruning**: where individual weights with a small saliency (or the least weight magnitude) are removed no matter which part of the neural network they are located in.

- **Structured pruning**: this approach prunes heads or layers.

However, the pruning process has to be compatible with modern GPUs.

Most libraries such as Torch or TensorFlow come with this capability. We will describe how it is possible to prune a model using Torch. There are many different methods to use in pruning (magnitude-based or mutual information-based). One of the simplest ones to understand and implement is the L1 pruning method. This method takes the weights of each layer and zeros out the ones with the lowest L1-norm. You can also specify what percentage of your weights must be converted to zero after pruning. In order to make this example more understandable and show its impact on the model, we'll use the text representation example from *Chapter 7, Text Representation*. We will prune the model and see how it performs after pruning:

1. We will use the Roberta model. You can load the model using the following code:

   ```
   from sentence_transformers import SentenceTransformer
   distilroberta = SentenceTransformer('stsb-distilroberta-
   base-v2')
   ```

2. You will also need to load metrics and datasets for evaluation:

   ```
   from datasets import load_metric, load_dataset
   stsb_metric = load_metric('glue', 'stsb')
   ```

```
stsb = load_dataset('glue', 'stsb')
mrpc_metric = load_metric('glue', 'mrpc')
mrpc = load_dataset('glue','mrpc')
```

3. In order to evaluate the model, just like in *Chapter 7, Text Representation*, you can use the following function:

```
import math
import tensorflow as tf
def roberta_sts_benchmark(batch):
    sts_encode1 = tf.nn.l2_normalize(
                    distilroberta.encode(batch['sentence1']),
                    axis=1)
    sts_encode2 = tf.nn.l2_normalize(
        distilroberta.encode(batch['sentence2']), axis=1)
    cosine_similarities = tf.reduce_sum(
        tf.multiply(sts_encode1, sts_encode2), axis=1)
    clip_cosine_similarities = tf.clip_by_value(cosine_
similarities,-1.0,1.0)
    scores = 1.0 -\
                tf.acos(clip_cosine_similarities) / math.pi
return scores
```

4. And of course, it is required to set labels:

```
references = stsb['validation'][:]['label']
```

5. And to run the base model with no changes in it:

```
distilroberta_results = roberta_sts_
benchmark(stsb['validation'])
```

6. After all these things are done, this is the step where we actually start to prune our model:

```
from torch.nn.utils import prune
pruner = prune.L1Unstructured(amount=0.2)
```

7. The previous code makes a pruning object using L1-norm pruning with 20% of the weights in each layer. To apply it to the model, you can use the following code:

```
state_dict = distilroberta.state_dict()
for key in state_dict.keys():
    if "weight" in key:
        state_dict[key] = pruner.prune(state_dict[key])
```

It will iteratively prune all layers that have weight in their name; in other words, we will prune all weight layers and not touch the layers that are biased. Of course, you can try that too for experimentation purposes.

8. And again, it is good to reload the state dictionary to the model:

```
distilroberta.load_state_dict(state_dict)
```

9. Now that we have done everything, we can test the new model:

```
distilroberta_results_p = roberta_sts_
benchmark(stsb['validation'])
```

10. In order to have a good visual representation of the results, you can use the following code:

```
import pandas as pd
pd.DataFrame({
"DistillRoberta":stsb_metric.
compute(predictions=distilroberta_results,
references=references),
"DistillRobertaPruned":stsb_metric.
compute(predictions=distilroberta_results_p,
references=references)
})
```

The following screenshot shows the results:

	DistillRoberta	DistillRobertaPruned
pearson	0.888461	0.849915
spearmanr	0.889246	0.849125

Figure 8.2 – Comparison between original and pruned models

But what you did is you eliminated 20% of all weights of the model, reduced its size and computation cost, and lost 4% in performance. However, this step can be combined with other techniques such as quantization, which is explored in the next subsection.

This type of pruning is applied to some of the weights in a layer; however, it is also possible to completely drop some parts or layers of transformer architectures, for example, it is possible to drop some of the attention heads and track the changes too.

There are also other types of pruning algorithms available in PyTorch, such as iterative and global pruning, which are worth trying.

Quantization

Quantization is a signal processing and communication term that is generally used to emphasize how much accuracy is presented by the data provided. More bits mean more accuracy and precision in terms of data resolution. For example, if you have a variable that is presented by 4 bits and you want to quantize it to 2 bits, it means you have to drop the accuracy of your resolution. With 4 bits, you can specify 16 different states, while with 2 bits you can distinguish 4 states. In other words, by reducing the resolution of your data from 4 to 2 bits, you are saving 50% more space and complexity.

Many popular libraries, such as TensorFlow, PyTorch, and MXNET, support mixed-precision operation. Recall the `fp16` parameter used in the `TrainingArguments` class in `chapter 05`. `fP16` increases computational efficiency since modern GPUs offer higher efficiency for reduced precision math, but the results are accumulated in `fP32`. Mixed-precision can reduce the memory usage required for training, which allows us to increase the batch size or model size.

Quantization can be applied to model weights to reduce their resolution and save computation time, memory, and storage. In this subsection, we will try to quantize the model that we pruned in the previous section:

1. In order to do so, you can use the following code to quantize your model in 8-bit integer representation instead of float:

```
import torch
distilroberta = torch.quantization.quantize_dynamic(
        model=distilroberta,
        qconfig_spec = {
        torch.nn.Linear :
        torch.quantization.default_dynamic_qconfig,
                          },
        dtype=torch.qint8)
```

2. Afterwards, you can get the evaluation results by using the following code:

```
distilroberta_results_pq = roberta_sts_
benchmark(stsb['validation'])
```

3. And as before, you can view the results:

```
pd.DataFrame({
"DistillRoberta":stsb_metric.
compute(predictions=distilroberta_results,
references=references),
"DistillRobertaPruned":stsb_metric.
compute(predictions=distilroberta_results_p,
references=references),
"DistillRobertaPrunedQINT8":stsb_metric.
compute(predictions=distilroberta_results_pq,
references=references)
})
```

The results can be seen as follows:

	DistillRoberta	DistillRobertaPruned	DistillRobertaPrunedQINT8
pearson	0.888461	0.849915	0.826784
spearmanr	0.889246	0.849125	0.824857

Figure 8.3 – Comparison between original, pruned, and quantized models

4. Until now, you just used a distilled model, pruned it, and then you quantized it to reduce its size and complexity. Let's see how much space you have saved by saving the model:

```
distilroberta.save("model_pq")
```

Use the following code in order to see the model size:

```
ls model_pq/0_Transformer/ -l --block-size=M | grep
pytorch_model.bin
-rw-r--r-- 1 root 191M May 23 14:53 pytorch_model.bin
```

As you can see, it is 191 MB. The initial size of the model was 313 MB, which means we managed to decrease the size of the model to 61% of its original size and just lost 6%-6.5% in terms of performance. Please note that the block-size parameter may fail on a Mac, and it is required to use -lh instead.

Up to this point, you have learned about pruning and quantization in terms of practical model preparation for industrial usage. However, you also gained information about the distillation process and how it can be useful. There are many other ways to perform pruning and quantization, which can be a good step to go in after reading this section. For more information and guides, you can take a look at **movement pruning** at `https://github.com/huggingface/block_movement_pruning`. This kind of pruning is a simple and deterministic first-order weight pruning approach. It uses the weight changes in training to find out which weights are more likely to be unused to have less effect on the result.

Working with efficient self-attention

Efficient approaches restrict the attention mechanism to get an effective transformer model because the computational and memory complexity of a transformer is mostly due to the self-attention mechanism. The attention mechanism scales quadratically with respect to the input sequence length. For short input, quadratic complexity may not be an issue. However, to process longer documents, we need to improve the attention mechanism that scales linearly with sequence length.

We can roughly group the efficient attention solutions into three types:

- Sparse attention with fixed patterns
- Learnable sparse patterns
- Low-rank factorization/kernel function

Let's begin with sparse attention based on a fixed pattern next.

Sparse attention with fixed patterns

Recall that the attention mechanism is made up of a query, key, and values as roughly formulated here:

$$Attention\ (Q, K, V)\ =\ Score\ (Q, K)\ .V$$

Here, the `Score` function, which is mostly softmax, performs QK^T multiplication that requires $O(n^2)$ memory and computational complexity since a token position attends to all other token positions in full self-attention mode to build its position embeddings. We repeat the same process for all token positions to get their embeddings, leading to a quadratic complexity problem. It is a very expensive way of learning, especially for long-context NLP problems. It is natural to ask the question do we need such a dense interaction or is there a cheaper way to do the calculations? Many researchers have addressed this problem and employed a variety of techniques to mitigate the complexity burden and to reduce the quadratic complexity of the self-attention mechanism. They have mostly made a trade-off between performance, computation, and memory, especially for long documents.

The simplest way of reducing complexity is to sparsify the full self-attention matrix or find another cheaper way to approximate full attention. Sparse attention patterns formulate how to connect/disconnect certain positions without disturbing the flow of information through layers, which helps the model to track long-term dependency and to build sentence-level encoding.

Full self-attention and sparse attention are depicted in *Figure 8.4* in that order, where the rows correspond to output positions and the columns are for the inputs. A full self-attention model would directly transfer the information between any two positions. On the other hand, in localized sliding window attention, which is sparse attention, as shown on the right of the figure, the empty cells mean that there is no interaction between the corresponding input-output position. The sparse model in the figure is based on fixed patterns that are certain manually designed rules. More specifically, it is localized sliding window attention that was one of the first proposed methods, also known as the local-based fixed pattern approach. The assumption behind it is that useful information is located in each position neighbor. Each query token attends to window/2 key tokens to the left and window/2 key tokens to the right of that position. In the following example, the window size is selected as 4. This rule applies to every layer in the transformer in the same way. In some studies, the window size is increased as it moves towards the layers further.

The following figure simply depicts the difference between full and sparse attention:

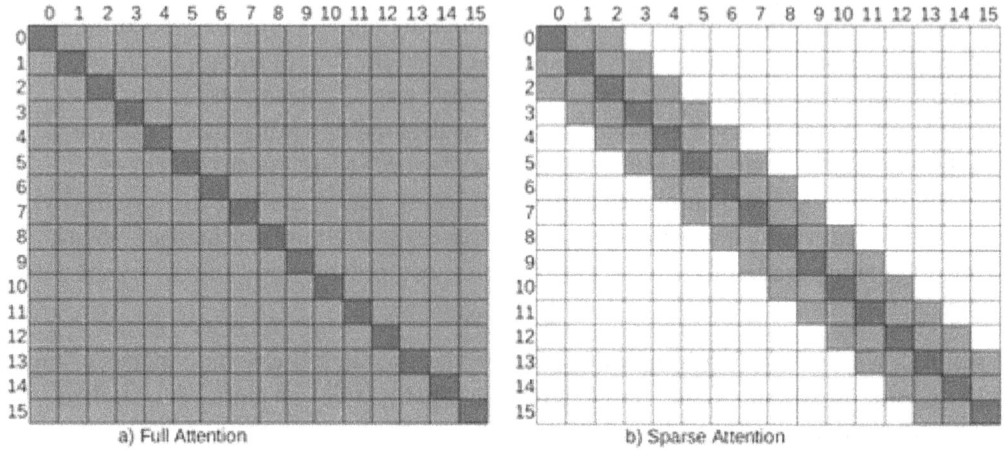

Figure 8.4 – Full attention versus sparse attention

In sparse mode, the information is transmitted through connected nodes (non-empty cells) in the model. For example, the output position 7 of the sparse attention matrix cannot directly attend to the input position 3 (please see the sparse matrix at the right of *Figure 8.4*) since the cell (7,3) is seen as empty. However, position 7 indirectly attends to position 3 via the token position 5, that is (7->5, 5->3 => 7->3). The figure also illustrates that while the full self-attention incurs n^2 number of active cells (vertex), the sparse model does roughly 5×n.

Another important type is global attention. A few selected tokens or a few injected tokens are used as global attention that can attend to all other positions and be attended by them. Hence, the maximum path distance between any two token positions is equal to 2. Suppose we have a sentence *[GLB, the, cat, is, very, sad]* where **Global (GLB)** is an injected global token and the window size is 2, which means a token can attend to only its immediate left-right tokens and to GLB as well. There is no direct interaction from *cat* to *sad*. But we can follow *cat-> GLB, GLB-> sad* interactions, which creates a hyperlink through the GLB token. The global tokens can be selected from existing tokens or added like *(CLS)*. As shown in the following screenshot, the first two token positions are selected as global tokens:

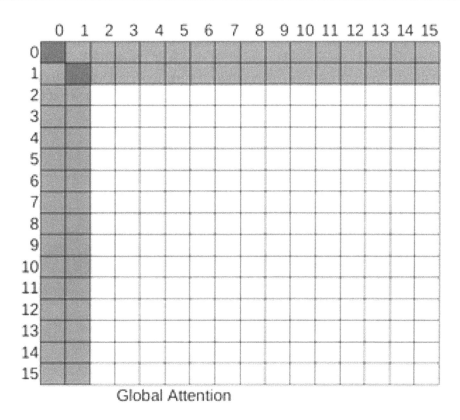

Figure 8.5 – Global attention

By the way, these global tokens don't have to be at the beginning of the sentence either. For example, the longformer model randomly selects global tokens in addition to the first two tokens..

There are four more widely seen patterns. **Random attention** (the first matrix in *Figure 8.6*) is used to ease the flow of information by randomly selecting from existing tokens. But most of the time, we employ random attention as part of a **combined pattern** (the bottom-left matrix) that consists of a combination of other models. **Dilated attention** is similar to the sliding window, but some gaps are put in the window as shown at the top right of *Figure 8.6*:

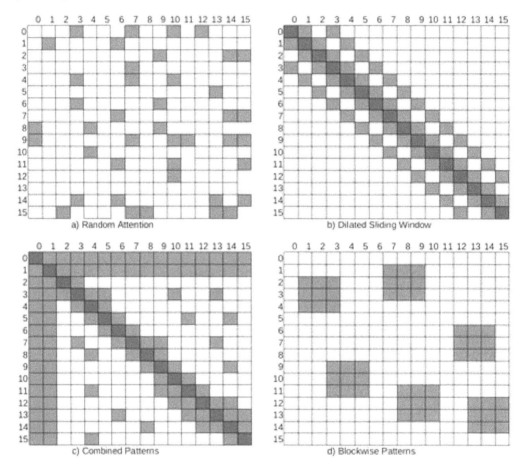

Figure 8.6 – Random, Dilated, Combined, and Blockwise

The **Blockwise Pattern** (the bottom right of *Figure 8.6*) provides a basis for other patterns. It chunks the tokens into a fixed number of blocks, which is especially useful for long-context problems. For example, when a 4,096x4,096 attention matrix is chunked using a block size of 512, then 8 (512x512) query blocks and key blocks are formed. Many efficient models such as BigBird and Reformer mostly chunk tokens into blocks to reduce the complexity.

It is important to note that the proposed patterns must be supported by the accelerators and the libraries. Some attention patterns such as dilated patterns require a special matrix multiplication that is not directly supported in current deep learning libraries such as PyTorch or TensorFlow as of writing this chapter.

We are ready to run some experiments for efficient transformers. We will proceed with models that are supported by the Transformers library and that have checkpoints on the HuggingFace platform. **Longformer** is one of the models that use sparse attention. It uses a combination of a sliding window and global attention. It supports dilated sliding window attention as well:

1. Before we start, we need to install the py3nvml package for benchmarking. Please recall that we already discussed how to apply benchmarking in *Chapter 2, A Hands-On Introduction to the Subject*:

    ```
    !pip install py3nvml
    ```

2. We also need to check our devices to ensure that there is no running process:

    ```
    !nvidia-smi
    ```

    The output is as follows:

```
Sat May 22 13:43:18 2021
+-------------------------------------------------------------------------+
| NVIDIA-SMI 465.19.01    Driver Version: 460.32.03    CUDA Version: 11.2  |
|-------------------------------+----------------------+------------------+
| GPU  Name        Persistence-M| Bus-Id        Disp.A | Volatile Uncorr. ECC |
| Fan  Temp  Perf  Pwr:Usage/Cap|         Memory-Usage | GPU-Util  Compute M. |
|                               |                      |               MIG M. |
|===============================+======================+==================|
|   0  Tesla P100-PCIE...  Off  | 00000000:00:04.0 Off |                0 |
| N/A   36C    P0    26W / 250W |      0MiB / 16280MiB |      0%    Default |
|                               |                      |              N/A |
+-------------------------------+----------------------+------------------+

+-------------------------------------------------------------------------+
| Processes:                                                              |
|  GPU   GI   CI        PID   Type   Process name            GPU Memory |
|        ID   ID                                             Usage      |
|=========================================================================|
|  No running processes found                                            |
+-------------------------------------------------------------------------+
```

Figure 8.7 – GPU usage

3. Currently, the Longformer author has shared a couple of checkpoints. The following code snippet loads the Longformer checkpoint `allenai/longformer-base-4096` and processes a long text:

```
from transformers import LongformerTokenizer,
LongformerForSequenceClassification
import torch
tokenizer = LongformerTokenizer.from_pretrained(
    'allenai/longformer-base-4096')
model=LongformerForSequenceClassification.from_
pretrained(
    'allenai/longformer-base-4096')
sequence= "hello "*4093
inputs = tokenizer(sequence, return_tensors="pt")
print("input shape: ",inputs.input_ids.shape)
outputs = model(**inputs)
```

4. The output is as follows:

```
input shape:  torch.Size([1, 4096])
```

As seen, Longformer can process a sequence up to the length of 4096. When we pass a sequence whose length is more than 4096, which is the limit, you will get the error `IndexError: index out of range in self`.

Longformer's default `attention_window` is 512, which is the size of the attention window around each token. With the following code, we instantiate two Longformer configuration objects, where the first one is the default Longformer, and the second is a lighter one where we set the window size to a smaller value such as 4 so that the model becomes lighter:

1. Please pay attention to the following examples. We will always call `XformerConfig.from_pretrained()`. This call does not download the actual weights of the model checkpoint, instead only downloading the configuration from the HuggingFace Hub. Throughout this section, since we will not fine-tune, we only need the configuration:

```
from transformers import LongformerConfig, \
PyTorchBenchmark, PyTorchBenchmarkArguments
config_longformer=LongformerConfig.from_pretrained(
    "allenai/longformer-base-4096")
config_longformer_window4=LongformerConfig.from_
```

```
pretrained(
    "allenai/longformer-base-4096",
    attention_window=4)
```

2. With these configuration instances, you can train your Longformer language model with your own datasets passing the configuration object to the Longformer model as follows:

```
from transformers import LongformerModel
model = LongformerModel(config_longformer)
```

Other than training a Longformer model, you can also fine-tune a trained checkpoint to a downstream task. To do so, you can continue by applying the code as shown in Chapter 03 for language model training and Chapter 05-06 for fine-tuning.

3. We will now compare the time and memory performance of these two configurations with various lengths of input [128, 256, 512, 1024, 2048, 4096] by utilizing PyTorchBenchmark as follows:

```
sequence_lengths=[128,256,512,1024,2048,4096]
models=["config_longformer","config_longformer_window4"]
configs=[eval(m) for m in models]
benchmark_args = PyTorchBenchmarkArguments(
    sequence_lengths= sequence_lengths,
    batch_sizes=[1],
    models= models)
benchmark = PyTorchBenchmark(
    configs=configs,
    args=benchmark_args)
results = benchmark.run()
```

4. The output is the following:

```
> 1 / 2
  2 / 2
```

```
====================        INFERENCE · SPEED · RESULT        ====================
- - - - - - - - - - - - - - - - - - - - - - - - - - - - - - - - - - - - - - - - -
          Model Name          Batch Size    Seq Length    Time in s
- - - - - - - - - - - - - - - - - - - - - - - - - - - - - - - - - - - - - - - - -
        config_longformer          1            128          0.036
        config_longformer          1            256          0.036
        config_longformer          1            512          0.036
        config_longformer          1           1024          0.064
        config_longformer          1           2048          0.117
        config_longformer          1           4096          0.226
   config_longformer_window4       1            128          0.019
   config_longformer_window4       1            256          0.022
   config_longformer_window4       1            512          0.028
   config_longformer_window4       1           1024          0.044
   config_longformer_window4       1           2048          0.074
   config_longformer_window4       1           4096          0.136
- - - - - - - - - - - - - - - - - - - - - - - - - - - - - - - - - - - - - - - - -


====================        INFERENCE · MEMORY · RESULT        ====================
- - - - - - - - - - - - - - - - - - - - - - - - - - - - - - - - - - - - - - - - -
          Model Name          Batch Size    Seq Length    Memory in MB
- - - - - - - - - - - - - - - - - - - - - - - - - - - - - - - - - - - - - - - - -
        config_longformer          1            128          1595
        config_longformer          1            256          1595
        config_longformer          1            512          1595
        config_longformer          1           1024          1679
        config_longformer          1           2048          1793
        config_longformer          1           4096          2089
   config_longformer_window4       1            128          1525
   config_longformer_window4       1            256          1527
   config_longformer_window4       1            512          1541
   config_longformer_window4       1           1024          1561
   config_longformer_window4       1           2048          1643
   config_longformer_window4       1           4096          1763
- - - - - - - - - - - - - - - - - - - - - - - - - - - - - - - - - - - - - - - - -
```

Figure 8.8 – Benchmark results

Some hints for PyTorchBenchmarkArguments: if you like to see the performance for training as well as inference, you should set the argument training to True (the default is False). You also may want to see your current environment information. You can do so by setting no_env_print to False; the default is True.

Let's visualize the performance to be more interpretable. To do so, we define a `plotMe()` function since we will need that function for further experiments as well. The function plots the inference performance in terms of both running time complexity by default or memory footprint properl:.

1.   Here is the function definition:

```python
import matplotlib.pyplot as plt
def plotMe(results,title="Time"):
    plt.figure(figsize=(8,8))
    fmts= ["rs--","go--","b+-","c-o"]
    q=results.memory_inference_result
    if title=="Time":
        q=results.time_inference_result
    models=list(q.keys())
    seq=list(q[models[0]]['result'][1].keys())
    models_perf=[list(q[m]['result'][1].values()) \
        for m in models]
    plt.xlabel('Sequence Length')
    plt.ylabel(title)
    plt.title('Inference Result')
    for perf,fmt in zip(models_perf,fmts):
        plt.plot(seq, perf,fmt)
    plt.legend(models)
    plt.show()
```

2.   Let's see the computational performance of two Longformer configurations, as follows:

```python
plotMe(results)
```

This plots the following chart:

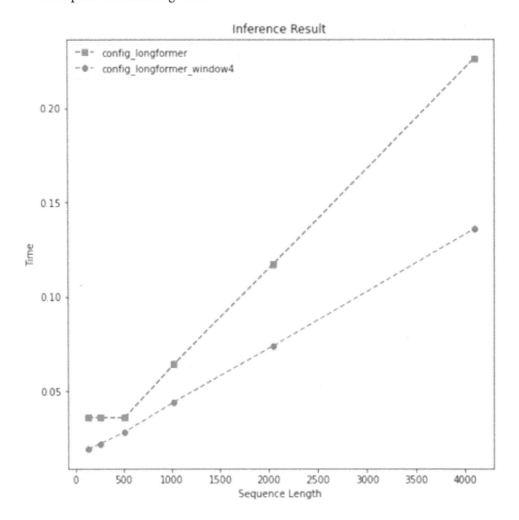

Figure 8.9 – Speed performance over sequence length (Longformer)

In this and the next examples, we see the main differentiation between a heavy model and a light model starting from length 512. The preceding figure shows the lighter Longformer model in green (the one with a window length of 4) performs better in terms of time complexity as expected. We also see that two Longformer models process the input with linear time complexity.

3. Let's evaluate these two models in terms of memory performance:

    ```
    plotMe(results, "Memory")
    ```

This plots the following:

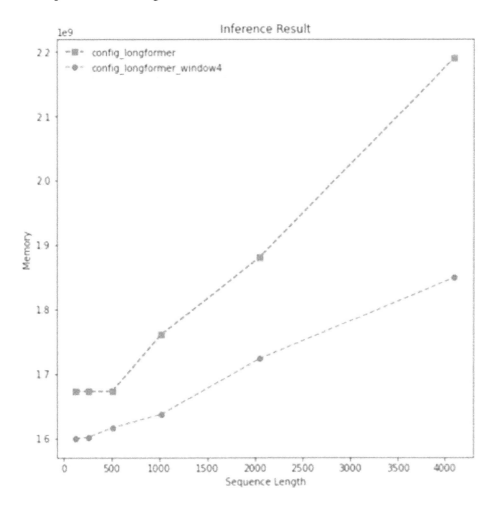

Figure 8.10 – Memory performance over sequence length (Longformer)

Again, up to length 512, there is no substantial differentiation. For the rest, we see a similar memory performance to the time performance. It is clear to say that the memory complexity of the Longformer self-attention is linear. On the other hand, let me bring to your attention that we are not saying anything about model task performance yet.

Thanks to the `PyTorchBenchmark` script, we have cross-checked these models. This script is very useful when we choose which configuration the language model should be trained with. It will be vital before starting the real language model training and fine-tuning.

Another best-performing model exploiting sparse attention is BigBird (Zohen et al. 2020). The authors claimed that their sparse attention mechanism (they called it a generalized attention mechanism) preserves all the functionality of the full self-attention mechanism of vanilla transformers in linear time. The authors treated the attention matrix as a directed graph so that they leveraged graph theory algorithms. They took inspiration from the graph sparsification algorithm, which approximates a given graph G by graph G' with fewer edges or vertices.

BigBird is a block-wise attention model and can handle sequences up to a length of 4096. It first blockifies the attention pattern by packing queries and keys together and then defines attention on these blocks. They utilize random, sliding window, and global attention.

4. Let's load and use the BigBird model checkpoint configuration just like the Longformer transformer model. There are a couple of BigBird checkpoints shared by the developers in the HuggingFace Hub. We select the original BigBird model, google/bigbird-roberta-base, which is warm started from a RoBERTa checkpoint. Once again, we're not downloading the model checkpoint weights but the configuration instead. The BigBirdConfig implementation allows us to compare full self-attention and sparse attention. Thus, we can observe and check whether the sparsification will reduce the full-attention O(n^2) complexity to a lower level. Once again, up to a length of 512, we do not clearly observe quadratic complexity. We can see the complexity from this level on. Setting the attention type to original-full will give us a full self-attention model. For comparison, we created two types of configurations: the first one is BigBird's original sparse approach, the second is a model that uses the full self-attention model.

5. We call them sparseBird and fullBird in order as follows:

```
from transformers import BigBirdConfig
# Default Bird with num_random_blocks=3, block_size=64
sparseBird = BigBirdConfig.from_pretrained(
    "google/bigbird-roberta-base")
fullBird = BigBirdConfig.from_pretrained(
    "google/bigbird-roberta-base",
    attention_type="original_full")
```

6. Please notice that for smaller sequence lengths up to 512, the BigBird model works as full self-attention mode due to block-size and sequence-length inconsistency:

```
sequence_lengths=[256,512,1024,2048, 3072, 4096]
models=["sparseBird","fullBird"]
```

```
configs=[eval(m) for m in models]
benchmark_args = PyTorchBenchmarkArguments(
    sequence_lengths=sequence_lengths,
    batch_sizes=[1],
    models=models)
benchmark = PyTorchBenchmark(
    configs=configs,
    args=benchmark_args)
results = benchmark.run()
```

The output is as follows:

```
================     INFERENCE · SPEED · RESULT     ====================
- - - - - - - - - - - - - - - - - - - - - - - - - - - - - - - - - - - - - -
    Model Name          Batch Size      Seq Length       Time in s
- - - - - - - - - - - - - - - - - - - - - - - - - - - - - - - - - - - - - -
    sparseBird              1               256            0.014
    sparseBird              1               512            0.029
    sparseBird              1               1024           0.112
    sparseBird              1               2048           0.288
    sparseBird              1               3072           0.217
    sparseBird              1               4096           0.279
    fullBird                1               256            0.014
    fullBird                1               512            0.024
    fullBird                1               1024           0.049
    fullBird                1               2048           0.123
    fullBird                1               3072           0.232
    fullBird                1               4096           0.348
- - - - - - - - - - - - - - - - - - - - - - - - - - - - - - - - - - - - - -

================     INFERENCE · MEMORY · RESULT     ====================
- - - - - - - - - - - - - - - - - - - - - - - - - - - - - - - - - - - - - -
    Model Name          Batch Size      Seq Length      Memory in MB
- - - - - - - - - - - - - - - - - - - - - - - - - - - - - - - - - - - - - -
    sparseBird              1               256            1495
    sparseBird              1               512            1571
    sparseBird              1               1024           1777
    sparseBird              1               2048           2195
    sparseBird              1               3072           2591
    sparseBird              1               4096           2969
    fullBird                1               256            1495
    fullBird                1               512            1571
    fullBird                1               1024           1801
    fullBird                1               2048           2257
    fullBird                1               3072           2955
    fullBird                1               4096           3835
- - - - - - - - - - - - - - - - - - - - - - - - - - - - - - - - - - - - - -
```

Figure 8.11 – Benchmark results (BigBird)

7. Again, we plot the time performance as follows:

```
plotMe(results)
```

This plots the following:

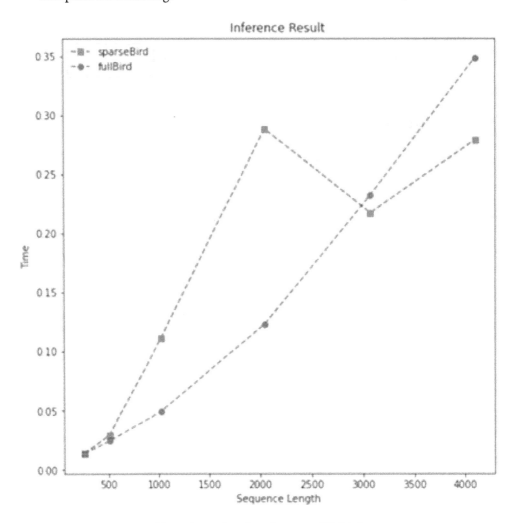

Figure 8.12 – Speed performance (BigBird)

To a certain extent, the full self-attention model performs better than a sparse model. However, we can observe the quadratic time complexity for fullBird. Hence, after a certain point, we also see that the sparse attention model abruptly outperforms it, when coming to an end.

8. Let's check the memory complexity as follows:

```
plotMe(results, "Memory")
```

Here is the output:

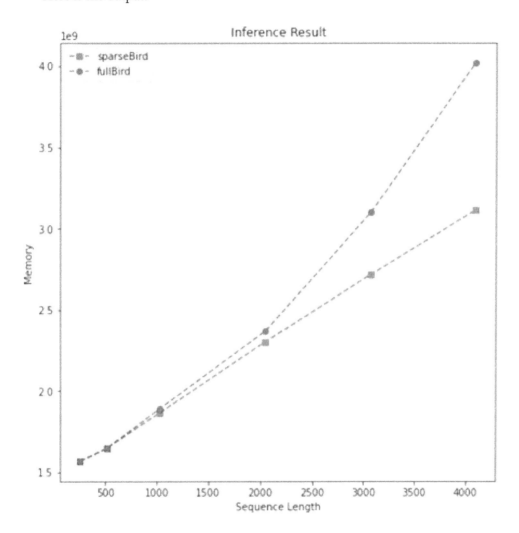

Figure 8.13 – Memory performance (BigBird)

In the preceding figure, we can clearly see linear and quadratic memory complexity. Once again, up to a certain point (a length of 2,000 in this example), we cannot speak of a clear distinction.

Next, let's discuss learnable patterns and work with models that can process longer input.

Learnable patterns

Learning-based patterns are the alternatives to fixed (predefined) patterns. These approaches extract the patterns in an unsupervised data-driven fashion. They leverage some techniques to measure the similarity between the queries and the keys to properly cluster them. This transformer family learns first how to cluster the tokens and then restrict the interaction to get an optimum view of the attention matrix.

Now, we will do some experiments with Reformer as one of the important efficient models based on learnable patterns. Before that, let's address what the Reformer model contributes to the NLP field, as follows:

1. It employs **Local Self Attention** (**LSA**) that cuts the input into **n** chunks to reduce the complexity bottleneck. But this cutting process makes the boundary token unable to attend to its immediate neighbors. For example, in the chunks [a,b,c] and [d,e,f], the token d cannot attend to its immediate context c. As a remedy, Reformer augments each chunk with the parameters that control the number of previous neighboring chunks.

2. The most important contribution of Reformer is to leverage the **Locality Sensitive Hashing** (**LSH**) function, which assigns the same value to similar query vectors. Attention could be approximated by only comparing the most similar vectors, which helps us reduce the dimensionality and then sparsify the matrix. It is a safe operation since the softmax function is highly dominated by large values and can ignore dissimilar vectors. Additionally, instead of finding the relevant keys to a given query, only similar queries are found and bucked. That is, the position of a query can only attend to the positions of other queries to which it has a high cosine similarity.

3. To reduce the memory footprint, Reformer uses reversible residual layers, which avoids the need to store the activations of all the layers to be reused for backpropagation, following the **Reversible Residual Network** (**RevNet**), because the activations of any layer can be recovered from the activation of the following layer.

    It is important to note that the Reformer model and many other efficient transformers are criticized as, in practice, they are only more efficient than the vanilla transformer when the input length is very long (*REF: Efficient Transformers: A Survey, Yi Tay, Mostafa Dehghani, Dara Bahri, Donald Metzler*). We made similar observations in our earlier experiments (please see the BigBird and Longformer experiment) .

4. Now we will conduct some experiments with Reformer. Thanks to the HuggingFace community again, the Transformers library provides us with Reformer implementation and its pre-trained checkpoints. We will load the configuration of the original checkpoint `google/reformer-enwik8` and also tweak some settings to work in full self-attention mode. When we set `lsh_attn_chunk_length` and `local_attn_chunk_length` to `16384`, which is the maximum length that Reformer can process, the Reformer instance will have no chance of local optimization and will automatically work like a vanilla transformer with full attention. We call it `fullReformer`. As for the original Reformer, we instantiate it with default parameters from the original checkpoint and call it `sparseReformer` as follows:

```
from transformers import ReformerConfig
fullReformer = ReformerConfig\
    .from_pretrained("google/reformer-enwik8",
        lsh_attn_chunk_length=16384,
        local_attn_chunk_length=16384)
sparseReformer = ReformerConfig\
    .from_pretrained("google/reformer-enwik8")
sequence_lengths=[256, 512, 1024, 2048, 4096, 8192,
12000]
models=["fullReformer","sparseReformer"]
configs=[eval(e) for e in models]
```

Please notice that the Reformer model can process sequences up to a length of `16384`. But for the full self-attention mode, due to the accelerator capacity of our environment, the attention matrix does not fit on GPU, and we get a CUDA out of memory warning. Hence, we set the max length as `12000`. If your environment is suitable, you can increase it.

5. Let's run the benchmark experiments as follows:

```
benchmark_args = PyTorchBenchmarkArguments(
    sequence_lengths=sequence_lengths,
    batch_sizes=[1],
    models=models)
benchmark = PyTorchBenchmark(
    configs=configs,
    args=benchmark_args)
result = benchmark.run()
```

The output is as follows:

```
=====================    INFERENCE · SPEED · RESULT    =====================
· · · · · · · · · · · · · · · · · · · · · · · · · · · · · · · · · · · · · · · · ·
        Model Name            Batch Size      Seq Length      Time in s
· · · · · · · · · · · · · · · · · · · · · · · · · · · · · · · · · · · · · · · · ·
        fullReformer              1              256           0.024
        fullReformer              1              512           0.036
        fullReformer              1             1024           0.075
        fullReformer              1             2048           0.196
        fullReformer              1             4096           0.529
        fullReformer              1             8192           1.722
        fullReformer              1            12000           3.443
      sparseReformer              1              256           0.026
      sparseReformer              1              512           0.049
      sparseReformer              1             1024           0.084
      sparseReformer              1             2048           0.165
      sparseReformer              1             4096           0.296
      sparseReformer              1             8192           0.576
      sparseReformer              1            12000           0.869
· · · · · · · · · · · · · · · · · · · · · · · · · · · · · · · · · · · · · · · · ·


=====================    INFERENCE · MEMORY · RESULT    =====================
· · · · · · · · · · · · · · · · · · · · · · · · · · · · · · · · · · · · · · · · ·
        Model Name            Batch Size      Seq Length     Memory in MB
· · · · · · · · · · · · · · · · · · · · · · · · · · · · · · · · · · · · · · · · ·
        fullReformer              1              256           1511
        fullReformer              1              512           1551
        fullReformer              1             1024           1671
        fullReformer              1             2048           2115
        fullReformer              1             4096           3791
        fullReformer              1             8192           8415
        fullReformer              1            12000          16191
      sparseReformer              1              256           1509
      sparseReformer              1              512           1705
      sparseReformer              1             1024           1885
      sparseReformer              1             2048           2339
      sparseReformer              1             4096           3143
      sparseReformer              1             8192           4803
      sparseReformer              1            12000           6367
· · · · · · · · · · · · · · · · · · · · · · · · · · · · · · · · · · · · · · · · ·
```

Figure 8.14 – Benchmark results

6. Let's visualize the time performance result as follows:

```
plotMe(result)
```

The output is the following:

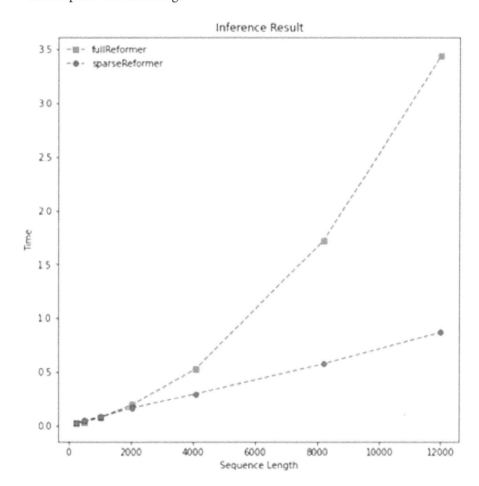

Figure 8.15 – Speed performance (Reformer)

7. We can see the linear and quadratic complexity of the models. We observe similar characteristics for the memory footprint by running the following line:

```
plotMe(result,"Memory Footprint")
```

It plots the following:

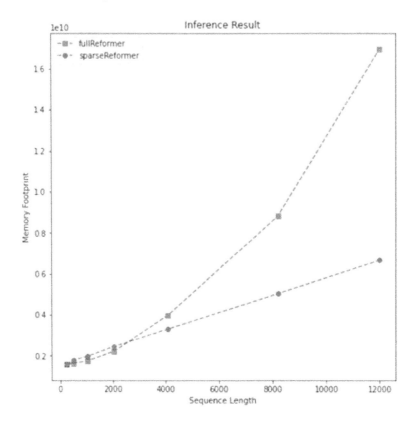

Figure 8.16 – Memory usage (Reformer)

Just as expected, Reformer with sparse attention produces a lightweight model. However, as was said before, we have difficulty observing the quadratic/linear complexity up to a certain length. As all these experiments indicate, efficient transformers can mitigate the time and memory complexity for longer text. What about task performance? How accurate would they be for classification or summarization tasks? To answer this, we will either start an experiment or have a look at the performance reports in the relevant articles of the models. For the experiment, you can repeat the code in chapter 04 and chapter 05 by instantiating an efficient model instead of a vanilla transformer. And you can track model performance and optimize it by using model tracking tools that we will discuss in detail in *Chapter 11, Attention Visualization and Experiment Tracking*.

Low-rank factorization, kernel methods, and other approaches

The latest trend of the efficient model is to leverage low-rank approximations of the full self-attention matrix. These models are considered to be the lightest since they can reduce the self-attention complexity from $O(n2)$ to $O(n)$ in both computational time and memory footprint. Choosing a very small projection dimension k, such that $k << n$, then the memory and space complexity is highly reduced. Linformer and Synthesizer are the models that efficiently approximate the full attention with a low-rank factorization. They decompose the dot-product $N \times N$ attention of the original transformer through linear projections.

Kernel attention is another method family that we have seen lately to improve efficiency by viewing the attention mechanism through kernelization. A kernel is a function that takes two vectors as arguments and returns the product of their projection with a feature map. It enables us to operate in high-dimensional feature space without even computing the coordinate of the data in that high-dimensional space, because computations within that space become more expensive. This is when the kernel trick comes into play. The efficient models based on kernelization enable us to re-write the self-attention mechanism to avoid explicitly computing the N×N matrix. In machine learning, the algorithm we hear the most about kernel methods is Support Vector Machines, where the radial basis function kernel or polynomial kernel are widely used, especially for nonlinearity. For transformers, the most notable examples are **Performer** and **Linear Transformers**.

Summary

The importance of this chapter is that we have learned how to mitigate the burden of running large models under limited computational capacity. We first discussed and implemented how to make efficient models out of trained models using distillation, pruning, and quantization. It is important to pre-train a smaller general-purpose language model such as DistilBERT. Such light models can then be fine-tuned with good performance on a wide variety of problems compared to their non-distilled counterparts.

Second, we have gained knowledge about efficient sparse transformers that replace the full self-attention matrix with a sparse one using approximation techniques such as Linformer, BigBird, Performer, and so on. We have seen how they perform on various benchmarks such as computational complexity and memory complexity. The examples showed us these approaches are able to reduce the quadratic complexity to linear complexity without sacrificing the performance.

In the next chapter, we will discuss other important topics: cross-lingual/multi-lingual models.

References

- Sanh, V., Debut, L., Chaumond, J., & Wolf, T. (2019). *DistilBERT, a distilled version of BERT: smaller, faster, cheaper and lighter*. arXiv preprint arXiv:1910.01108.

- Choromanski, K., Likhosherstov, V., Dohan, D., Song, X., Gane, A., Sarlos, T., & Weller, A. (2020). *Rethinking attention with performers*. arXiv preprint arXiv:2009.14794.

- Wang, S., Li, B., Khabsa, M., Fang, H., & Ma, H. (2020). *Linformer: Self-attention with linear complexity*. arXiv preprint arXiv:2006.04768.

- Zaheer, M., Guruganesh, G., Dubey, A., Ainslie, J., Alberti, C., Ontanon, S., ... & Ahmed, A. (2020). *Big bird: Transformers for longer sequences*. arXiv preprint arXiv:2007.14062.

- Tay, Y., Dehghani, M., Bahri, D., & Metzler, D. (2020). *Efficient transformers: A survey*. arXiv preprint arXiv:2009.06732.

- Tay, Y., Bahri, D., Metzler, D., Juan, D. C., Zhao, Z., & Zheng, C. (2020). *Synthesizer: Rethinking self-attention in transformer models*. arXiv preprint arXiv:2005.00743.

- Kitaev, N., Kaiser, Ł., & Levskaya, A. (2020). *Reformer: The efficient transformer*. arXiv preprint arXiv:2001.04451.

- Fournier, Q., Caron, G. M., & Aloise, D. (2021). *A Practical Survey on Faster and Lighter Transformers*. arXiv preprint arXiv:2103.14636.

9
Cross-Lingual and Multilingual Language Modeling

Up to this point, you have learned a lot about transformer-based architectures, from encoder-only models to decoder-only models, from efficient transformers to long-context transformers. You also learned about semantic text representation based on a Siamese network. However, we discussed all these models in terms of monolingual problems. We assumed that these models just understand a single language and are not capable of having a general understanding of text, regardless of the language itself. In fact, some of these models have multilingual variants; **Multilingual Bidirectional Encoder Representations from Transformers (mBERT)**, **Multilingual Text-to-Text Transfer Transformer (mT5)**, and **Multilingual Bidirectional and Auto-Regressive Transformer (mBART)**, to name but a few. On the other hand, some models are specifically designed for multilingual purposes trained with cross-lingual objectives. For example, **Cross-lingual Language Model (XLM)** is such a method, and this will be described in detail in this chapter.

In this chapter, the concept of knowledge sharing between languages will be presented, and the impact of **Byte-Pair Encoding** (**BPE**) on the tokenization part is also another important subject to cover in order to achieve better input. Cross-lingual sentence similarity using the **Cross-Lingual Natural Language Inference** (**XNLI**) corpus will be detailed. Tasks such as cross-lingual classification and utilization of cross-lingual sentence representation for training on one language and testing on another one will be presented by concrete examples of real-life problems in **Natural Language Processing** (**NLP**), such as multilingual intent classification.

In short, you will learn the following topics in this chapter:

- Translation language modeling and cross-lingual knowledge sharing
- XLM and mBERT
- Cross-lingual similarity tasks
- Cross-lingual classification
- Cross-lingual zero-shot learning
- Fundamental limitations of multilingual models

Technical requirements

The code for this chapter is found in the repo at `https://github.com/PacktPublishing/Mastering-Transformers/tree/main/CH09`, which is in the GitHub repository for this book. We will be using Jupyter Notebook to run our coding exercises that require Python 3.6.0+, and the following packages will need to be installed:

- `tensorflow`
- `pytorch`
- `transformers >=4.00`
- `datasets`
- `sentence-transformers`
- `umap-learn`
- `openpyxl`

Check out the following link to see the Code in Action video:

`https://bit.ly/3zASz7M`

Translation language modeling and cross-lingual knowledge sharing

So far, you have learned about **Masked Language Modeling** (**MLM**) as a cloze task. However, language modeling using neural networks is divided into three categories based on the approach itself and its practical usage, as follows:

- MLM

- **Causal Language Modeling** (**CLM**)

- **Translation Language Modeling** (**TLM**)

It is also important to note that there are other pre-training approaches such as **Next Sentence Prediction** (**NSP**) and **Sentence Order Prediction** (**SOP**) too, but we just considered token-based language modeling. These three are the main approaches that are used in the literature. **MLM**, described and detailed in previous chapters, is a very close concept to a cloze task in language learning.

CLM is defined by predicting the next token, which is followed by some previous tokens. For example, if you see the following context, you can easily predict the next token:

<s> Transformers changed the natural language …

As you see, only the last token is masked, and the previous tokens are given to the model to predict that last one. This token would be *processing* and if the context with this token is given to you again, you might end it with an *"</s>"* token. In order to have good training on this approach, it is required to not mask the first token, because the model would have just a sentence start token to make a sentence out of it. This sentence can be anything! Here's an example:

<s> …

What would you predict out of this? It can be literally anything. To have better training and better results, it is required to give at least the first token, such as this:

<s> Transformers …

And the model is required to predict the *change*; after giving it *Transformers changed …* it is required to predict *the*, and so on. This approach is very similar to N-grams and **Long-Short-Term Memory** (**LSTM**)-based approaches because it is left-to-right modeling based on the probability $P(wn|wn-1, wn-2 ,…,w0)$ where wn is the token to be predicted and the rest is the tokens before it. The token with the maximum probability is the predicted one.

These are the objectives used for monolingual models. So, what can be done for cross-lingual models? The answer is **TLM**, which is very similar to MLM, with a few changes. Instead of giving a sentence from a single language, a sentence pair is given to a model in different languages, separated by a special token. The model is required to predict the masked tokens, which are randomly masked in any of these languages.

The following sentence pair is an example of such a task:

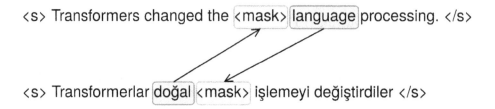

Figure 9.1 – Cross-lingual relation example between Turkish and English

Given these two masked sentences, the model is required to predict the missing tokens. In this task, on some occasions, the model has access to the tokens (for example, **doğal** and **language** respectively in the sentence pair in *Figure 9.1*) that are missing from one of the languages in the pair.

As another example, you can see the same pair from Persian and Turkish sentences. In the second sentence, the **değiştirdiler** token can be attended by the multiple tokens (one is masked) in the first sentence. In the following example, the word تغییری is missing but the meaning of **değiştirdiler** is تغییری دادند. :

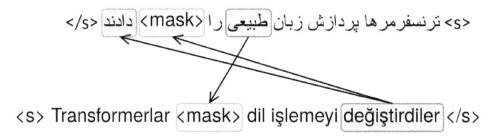

Figure 9.2 – Cross-lingual relation example between Persian and Turkish

Accordingly, a model can learn the mapping between these meanings. Just as with a translation model, our TLM must also learn these complexities between languages because **Machine Translation** (**MT**) is more than a token-to-token mapping.

XLM and mBERT

We have picked two models to explain in this section: mBERT and XLM. We selected these models because they correspond to the two best multilingual types as of writing this article. mBERT is a multilingual model trained on a different corpus of various languages using MLM modeling. It can operate separately for many languages. On the other hand, XLM is trained on different corpora using MLM, CLM, and TLM language modeling, and can solve cross-lingual tasks. For instance, it can measure the similarity of the sentences in two different languages by mapping them in a common vector space, which is not possible with mBERT.

mBERT

You are familiar with the BERT autoencoder model from *Chapter 3*, *Autoencoding Language Models*, and how to train it using MLM on a specified corpus. Imagine a case where a wide and huge corpus is provided not from a single language, but from 104 languages instead. Training on such a corpus would result in a multilingual version of BERT. However, training on such a wide variety of languages would increase the model size, and this is inevitable in the case of BERT. The vocabulary size would be increased and, accordingly, the size of the embedding layer would be larger because of more vocabulary.

Compared to a monolingual pre-trained BERT, this new version is capable of handling multiple languages inside a single model. However, the downside for this kind of modeling is that this model is not capable of mapping between languages. This means that the model, in the pre-training phase, does not learn anything about these mappings between semantic meanings of the tokens from different languages. In order to provide cross-lingual mapping and understanding for this model, it is necessary to train it on some of the cross-lingual supervised tasks, such as those available in the XNLI dataset.

Using this model is as easy as working with the models you have used in the previous chapters (see https://huggingface.co/bert-base-multilingual-uncased for more details). Here's the code you'll need to get started:

```
from transformers import pipeline
unmasker = pipeline('fill-mask', model='bert-base-
                    multilingual-uncased')
sentences = [
"Transformers changed the [MASK] language processing",
"Transformerlar [MASK] dil işlemeyi değiştirdiler",
"ترنسفرمرها پردازش زبان را [MASK] تغییری دادند"
]
```

```
for sentence in sentences:
    print(sentence)
    print(unmasker(sentence)[0]["sequence"])
    print("="*50)
```

The output will then be presented, as shown in the following code snippet:

```
Transformers changed the [MASK] language processing
transformers changed the english language processing
==================================================
Transformerlar [MASK] dil işlemeyi değiştirdiler
transformerlar bu dil islemeyi degistirdiler
==================================================
ترنسفرمرها پردازش زبان [MASK] را تغییری دادند
ترنسفرمرها پردازش زبانی را تغییری دادند
==================================================
```

As you can see, it can perform `fill-mask` for various languages.

XLM

Cross-lingual pre-training of language models, such as that shown with an XLM approach, is based on three different pre-training objectives. MLM, CLM, and TLM are used to pre-train the XLM model. The sequential order of this pre-training is performed using a shared BPE tokenizer between all languages. The reason that tokens are shared is that the shared tokens provide fewer tokens in the case of languages that have similar tokens or subwords, and on the other hand, these tokens can provide shared semantics in the pre-training process. For example, some tokens have remarkably similar writing and meaning across many languages, and accordingly, these tokens are shared by BPE for all. On the other hand, some tokens spelled the same in different languages can have different meanings—for example, *was* is shared in German and English contexts. Luckily, self-attention mechanisms help us to disambiguate the meaning of *was* using the surrounding context.

Another major improvement of this cross-lingual modeling is that it is also pre-trained on CLM, which makes it more reasonable for inferences where sentence prediction or completion is required. In other words, this model has an understanding of the languages and is capable of completing sentences, predicting missing tokens, and predicting missing tokens by using the other language source.

The following diagram shows the overall structure of cross-lingual modeling. You can read more at https://arxiv.org/pdf/1901.07291.pdf:

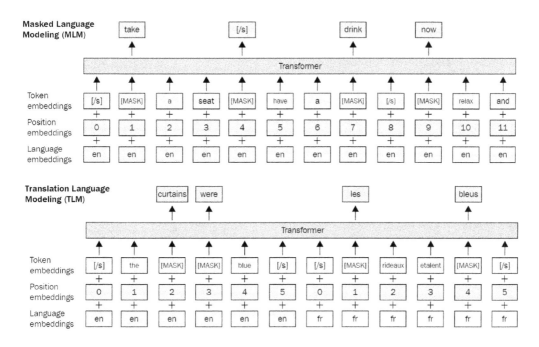

Figure 9.3 – MLM and TLM pre-training for cross-lingual modeling

A newer version of the XLM model is also released as **XLM-R**, which has minor changes in the training and corpus used. XLM-R is identical to the XLM model but is trained on more languages and a much bigger corpus. The **CommonCrawl** and **Wikipedia** corpus is aggregated, and the XLM-R is trained for MLM on it. However, the XNLI dataset is also used for TLM. The following diagram shows the amount of data used by XLM-R pre-training:

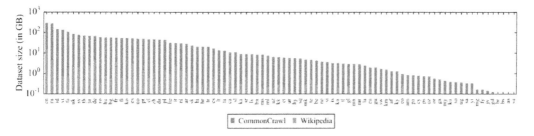

Figure 9.4 – Amount of data in gigabytes (GB) (log-scale)

There are many upsides and downsides when adding new languages for training data—for example, adding new languages may not always improve the overall model of **Natural Language Inference** (**NLI**). The **XNLI dataset** is usually used for multilingual and cross-lingual NLI. From previous chapters, you have seen the **Multi-Genre NLI** (**MNLI**) dataset for English; the XNLI dataset is almost identical to it but has more languages, and it also has sentence pairs. However, training only on this task is not enough, and it will not cover TLM pre-training. For TLM pre-training, much broader datasets such as the parallel corpus of **OPUS** (short for **Open Source Parallel Corpus**) are used. This dataset contains subtitles from different languages, aligned and cleaned, with the translations provided by many software sources such as Ubuntu, and so on.

The following screenshot shows OPUS (`https://opus.nlpl.eu/trac/`) and its components for searching and getting information about the dataset:

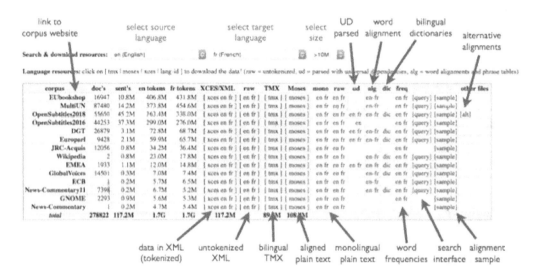

Figure 9.5 – OPUS

The steps for using cross-lingual models are described here:

1. Simple changes to the previous code can show you how XLM-R performs mask filling. First, you must change the model, as follows:

```
unmasker = pipeline('fill-mask', model='xlm-roberta-
base')
```

2. Afterward, you need to change the mask token from [MASK] to <mask>, which is a special token for XLM-R (or simply call tokenizer.mask_token). Here's the code to accomplish this:

```
sentences = [
"Transformers changed the <mask> language processing",
"Transformerlar <mask> dil işlemeyi değiştirdiler",
"ترن‌سفرمرها پردازش زبان <mask" را تغییری دادند"
]
```

3. Then, you can run the same code, as follows:

```
for sentence in sentences:
   print(sentence)
   print(unmasker(sentence)[0]["sequence"])
   print("="*50)
```

4. The results will appear, like so:

```
Transformers changed the <mask> language processing
Transformers changed the human language processing
==================================================
Transformerlar <mask> dil işlemeyi değiştirdiler
Transformerlar, dil işlemeyi değiştirdiler
==================================================
ترن‌سفرمرها پردازش زبان [MASK] را تغییری دادند
ترن‌سفرمرها پردازش زبانی را تغییری دادند
==================================================
```

5. But as you see from the Turkish and Persian examples, the model still made mistakes; for example, in the Persian text, it just added ی, and in the Turkish version, it added ,. For the English sentence, it added human, which is not what was expected. The sentences are not wrong, but not what we expected. However, this time, we have a cross-lingual model that is trained using TLM; so, let's use it by concatenating two sentences and giving the model some extra hints. Here we go:

```
print(unmasker("Transformers changed the natural language
processing. </s> Transformerlar <mask> dil işlemeyi
değiştirdiler.")[0]["sequence"])
```

6. The results will be shown, as follows:

```
Transformers changed the natural language processing.
Transformerlar doğal dil işlemeyi değiştirdiler.
```

7. That's it! The model has now made the right choice. Let's play with it a bit more and see how it performs, as follows:

```
print(unmasker("Earth is a great place to live in. </s>
زمین جای خوبی برای کردن است." ) [0]["sequence"])
```

Here is the result:

```
Earth is a great place to live in. زمین جای خوبی برای زندگی
کردن است.
```

Well done! So far, you have learned about multilingual and cross-lingual models such as mBERT and XLM. In the next section, you will learn how to use such models for multilingual text similarity. You will also see some use cases, such as multilingual plagiarism detection.

Cross-lingual similarity tasks

Cross-lingual models are capable of representing text in a unified form, where sentences are from different languages but those with close meaning are mapped to similar vectors in vector space. XLM-R, as was detailed in the previous section, is one of the successful models in this scope. Now, let's look at some applications on this.

Cross-lingual text similarity

In the following example, you will see how it is possible to use a cross-lingual language model pre-trained on the XNLI dataset to find similar texts from different languages. A use-case scenario is where a plagiarism detection system is required for this task. We will use sentences from the Azerbaijani language and see whether XLM-R finds similar sentences from English—if there are any. The sentences from both languages are identical. Here are the steps to take:

1. First, you need to load a model for this task, as follows:

```
from sentence_transformers import SentenceTransformer,
util
model = SentenceTransformer("stsb-xlm-r-multilingual")
```

2. Afterward, we assume that we have sentences ready in the form of two separate lists, as illustrated in the following code snippet:

```
azeri_sentences = ['Pişik çöldə oturur',
                   'Bir adam gitara çalır',
                   'Mən makaron sevirəm',
                   'Yeni film möhtəşəmdir',
                   'Pişik bağda oynayır',
                   'Bir qadın televizora baxır',
                   'Yeni film çox möhtəşəmdir',
                   'Pizzanı sevirsən?']
english_sentences = ['The cat sits outside',
                     'A man is playing guitar',
                     'I love pasta',
                     'The new movie is awesome',
                     'The cat plays in the garden',
                     'A woman watches TV',
                     'The new movie is so great',
                     'Do you like pizza?']
```

3. And the next step is to represent these sentences in vector space by using the XLM-R model. You can do this by simply using the encode function of the model, as follows:

```
azeri_representation = model.encode(azeri_sentences)
english_representation = \
model.encode(english_sentences)
```

4. At the final step, we will search for semantically similar sentences of the first language on the other language's representations, as follows:

```
results = []
for azeri_sentence, query in zip(azeri_sentences, azeri_
representation):
  id_, score = util.semantic_search(
        query,english_representation)[0][0].values()
  results.append({
      "azeri": azeri_sentence,
      "english": english_sentences[id_],
      "score": round(score, 4)
  })
```

5. In order to see a clear form of these results, you can use a pandas DataFrame, as follows:

```
import pandas as pd
pd.DataFrame(results)
```

And you will see the results with their matching score, as follows:

	azeri	english	score
0	Pişik çöldə oturur	The cat sits outside	0.5969
1	Bir adam gitara çalır	A man is playing guitar	0.9939
2	Mən makaron sevirəm	I love pasta	0.6878
3	Yeni film möhtəşəmdir	The new movie is so great	0.9757
4	Pişik bağda oynayır	A man is playing guitar	0.2695
5	Bir qadın televizora baxır	A woman watches TV	0.9946
6	Yeni film çox möhtəşəmdir	The new movie is so great	0.9797
7	Pizzanı sevirsən?	Do you like pizza?	0.9894

Figure 9.6 – Plagiarism detection results (XLM-R)

The model made mistakes in one case (row number *4*) if we accept the maximum scored sentence to be paraphrased or translated, but it is useful to have a threshold and accept values higher than it. We will show more comprehensive experimentation in the following sections.

On the other hand, there are alternative bi-encoders available too. Such approaches provide a pair encoding of two sentences and classify the result to train the model. In such cases, **Language-Agnostic BERT Sentence Embedding (LaBSE)** may be a good choice too, and it is available in the **sentence-transformers** library and in **TensorFlow Hub** too. LaBSE is a dual encoder based on Transformers, which is similar to Sentence-BERT, where two encoders that have the same parameters are combined with a loss function based on the dual similarity of two sentences.

Using the same example, you can change the model to LaBSE in a very simple way and rerun the previous code (*Step 1*), as follows:

```
model = SentenceTransformer("LaBSE")
```

The results are shown in the following screenshot:

	azeri	english	score
0	Pişik çöldə oturur	The cat sits outside	0.8686
1	Bir adam gitara çalır	A man is playing guitar	0.9143
2	Mən makaron sevirəm	I love pasta	0.8888
3	Yeni film möhtəşəmdir	The new movie is so great	0.9107
4	Pişik bağda oynayır	The cat sits outside	0.6761
5	Bir qadın televizora baxır	A woman watches TV	0.9359
6	Yeni film çox möhtəşəmdir	The new movie is so great	0.9258
7	Pizzanı sevirsən?	Do you like pizza?	0.9366

Figure 9.7 – Plagiarism detection results (LaBSE)

As you see, LaBSE performs better in this case, and the result in row number *4* is correct this time. LaBSE authors claim that it works very well in finding translations of sentences, but it is not so good at finding sentences that are not completely identical. For this purpose, it is a very useful tool for finding plagiarism in cases where a translation is used to steal intellectual material. However, there are many other factors that change the results too—for example, the resource size for the pre-trained model in each language and the nature of the language pairs is also important. For a reasonable comparison, we need a more comprehensive experiment, and we should consider many factors.

Visualizing cross-lingual textual similarity

Now, we will measure and visualize the degree of textual similarity between two sentences, one of which is a translation of the other. **Tatoeba** is a free collection of such sentences and translations, and it is part of the XTREME benchmark. The community aims to get high-quality sentence translation with the support of many participants. We'll now take the following steps:

1. We will get Russian and English sentences out of this collection. Make sure the following libraries are installed before you start working:

   ```
   !pip install sentence_transformers datasets transformers
   umap-learn
   ```

2. Load the sentence pairs, as follows:

   ```
   from datasets import load_dataset
   import pandas as pd
   data=load_dataset("xtreme","tatoeba.rus",
                     split="validation")
   pd.DataFrame(data)[["source_sentence","target_sentence"]]
   ```

   Let's look at the output, as follows:

	source_sentence	target_sentence
0	Я знаю много людей, у которых нет прав.\n	I know a lot of people who don't have driver's...
1	У меня много знакомых, которые не умеют играть...	I know a lot of people who don't know how to p...
2	Мой начальник отпустил меня сегодня пораньше.\n	My boss let me leave early today.\n
3	Я загорел на пляже.\n	I tanned myself on the beach.\n
4	Вы сегодня проверяли почту?\n	Have you checked your email today?\n
...	...	...
995	Что сказал врач?\n	What did the doctor say?\n
996	Я рад, что ты сегодня здесь.\n	I'm glad you're here today.\n
997	Фермеры пригнали в деревню пять волов, девять ...	The farmers had brought five oxen and nine cow...
998	Жужжание пчёл заставляет меня немного нервнича...	The buzzing of the bees makes me a little nerv...
999	С каждым годом они становились всё беднее.\n	From year to year they were growing poorer.\n

   1000 rows × 2 columns

   Figure 9.8 – Russian-English sentence pairs

3. First, we will take the first *K=30* sentence pairs for visualization, and later, we will run an experiment for the entire set. Now, we will encode them with sentence transformers that we already used for the previous example. Here is the execution of the code:

```
from sentence_transformers import SentenceTransformer
model = SentenceTransformer("stsb-xlm-r-multilingual")
K=30
q=data["source_sentence"][:K] + data["target_sentence"]
[:K]
emb=model.encode(q)
len(emb), len(emb[0])
Output: (60, 768)
```

4. We now have 60 vectors of length 768. We will reduce the dimensionality to 2 with **Uniform Manifold Approximation and Projection (UMAP)**, which we have already encountered in previous chapters. We visualized sentences that are translations of each other, marking them with the same color and code. We also drew a dashed line between them to make the link more obvious. The code is illustrated in the following snippet:

```
import matplotlib.pyplot as plt
import numpy as np
import umap
import pylab
X= umap.UMAP(n_components=2, random_state=42).fit_
transform(emb)
idx= np.arange(len(emb))
fig, ax = plt.subplots(figsize=(12, 12))
ax.set_facecolor('whitesmoke')
cm = pylab.get_cmap("prism")
colors = list(cm(1.0*i/K) for i in range(K))
for i in idx:
    if i<K:
        ax.annotate("RUS-"+str(i), # text
                    (X[i,0], X[i,1]), # coordinates
                    c=colors[i]) # color
        ax.plot((X[i,0],X[i+K,0]),(X[i,1],X[i+K,1]),"k:")
    else:
```

```
ax.annotate("EN-"+str(i%K),
             (X[i,0], X[i,1]),
             c=colors[i%K])
```

Here is the output of the preceding code:

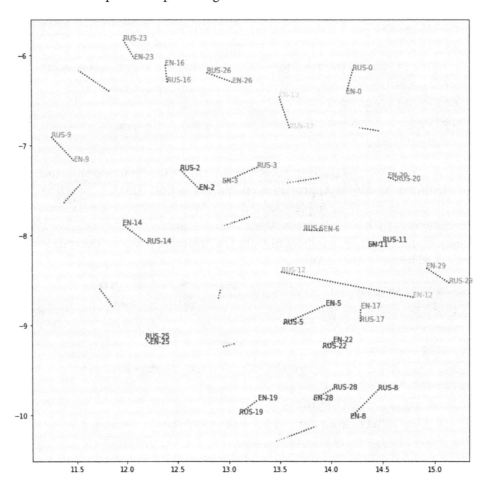

Figure 9.9 – Russian-English sentence similarity visualization

As we expected, most sentence pairs are located close to each other. Inevitably, some certain pairs (such as id 12) insist on not getting close.

5. For a comprehensive analysis, let's now measure the entire dataset. We encode all of the source and target sentences—1K pairs—as follows:

```
source_emb=model.encode(data["source_sentence"])
target_emb=model.encode(data["target_sentence"])
```

6. We calculate the cosine similarity between all pairs, save them in the `sims` variable, and plot a histogram, as follows:

```
from scipy import spatial
sims=[ 1 - spatial.distance.cosine(s,t) \
        for s,t in zip(source_emb, target_emb)]
plt.hist(sims, bins=100, range=(0.8,1))
plt.show()
```

Here is the output:

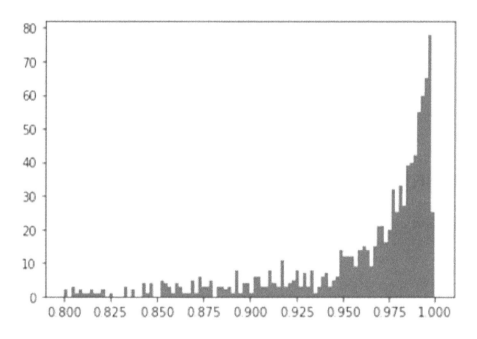

Figure 9.10 – Similarity histogram for the English and Russian sentence pairs

7. As can be seen, the scores are very close to 1. This is what we expect from a good cross-lingual model. The mean and standard deviation of all similarity measurements also support the cross-lingual model performance, as follows:

```
>>> np.mean(sims), np.std(sims)
(0.946, 0.082)
```

8. You can run the same code yourself for languages other than Russian. As you run it with **French** (`fra`), **Tamil** (`tam`), and so on, you will get the following resulting table. The table indicates that you will see in your experiment that the model works well in many languages but fails in others, such as **Afrikaans** or **Tamil**:

Language	Code	Mean	Std
French	`fra`	0.94	0.087
Afrikaans	`afr`	0.79	0.18
Arabic	`ara`	0.94	0.08
Korean	`kor`	0.92	0.11
Tamil	`tam`	0.77	0.19

Table 1 – Cross-lingual model performance for other languages

In this section, we applied cross-lingual models to measure similarity between different languages. In the next section, we'll make use of cross-lingual models in a supervised way.

Cross-lingual classification

So far, you have learned that cross-lingual models are capable of understanding different languages in semantic vector space where similar sentences, regardless of their language, are close in terms of vector distance. But how it is possible to use this capability in use cases where we have few samples available?

For example, you are trying to develop an intent classification for a chatbot in which there are few samples or no samples available for the second language; but for the first language—let's say English—you do have enough samples. In such cases, it is possible to freeze the cross-lingual model itself and just train a classifier for the task. A trained classifier can be tested on a second language instead of the language it is trained on.

In this section, you will learn how to train a cross-lingual model in English for text classification and test it in other languages. We have selected a very low-resource language known as **Khmer** (`https://en.wikipedia.org/wiki/Khmer_language`), which is spoken by 16 million people in Cambodia, Thailand, and Vietnam. It has few resources on the internet, and it is hard to find good datasets to train your model on it. However, we have access to a good **Internet Movie Database** (**IMDb**) sentiment dataset of movie reviews for sentiment analysis. We will use that dataset to find out how our model performs on the language it is not trained on.

The following diagram nicely depicts the kind of flow we will follow. The model is trained with train data on the left, and this model is applied to the test sets on the right. Please notice that MT and sentence-encoder mappings play a significant role in the flow:

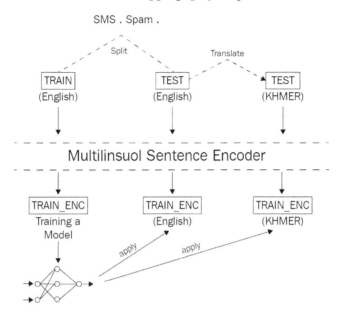

Figure 9.11 – Flow of cross-lingual classification

The required steps to load and train a model for cross-lingual testing are outlined here:

1. The first step is to load the dataset, as follows:

    ```
    from datasets import load_dataset
    sms_spam = load_dataset("imdb")
    ```

2. You need to shuffle the dataset to shuffle the samples before using them, as follows:

    ```
    imdb = imdb.shuffle()
    ```

3. The next step is to make a good test split out of this dataset, which is in the Khmer language. In order to do so, you can use a translation service such as Google Translate. First, you should save this dataset in Excel format, as follows:

    ```
    imdb_x = [x for x in imdb['train'][:1000]['text']]
    labels = [x for x in imdb['train'][:1000]['label']]
    import pandas as pd
    pd.DataFrame(imdb_x,
                columns=["text"]).to_excel(
    ```

```
                                            "imdb.xlsx",
                                         index=None)
```

4. Afterward, you can upload it to Google Translate and get the Khmer translation of this dataset (`https://translate.google.com/?sl=en&tl=km&op=docs`), as illustrated in the following screenshot:

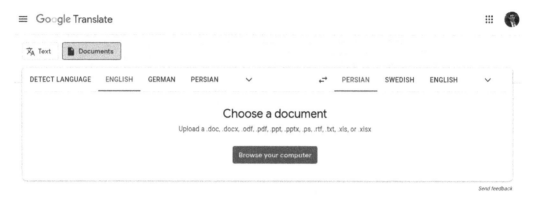

Figure 9.12 – Google document translator

5. After selecting and uploading the document, it will give you the translated version in Khmer, which you can copy and paste into an Excel file. It is also required to save it in Excel format again. The result would be an Excel document that is a translation of the original spam/ham English dataset. You can read it using pandas by running the following command:

```
pd.read_excel("KHMER.xlsx")
```

And the result will be seen, as follows:

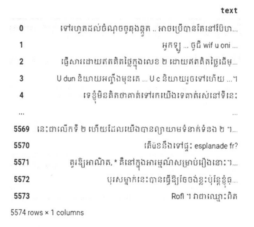

Figure 9.13 – IMDB dataset in KHMER language.

6. However, it is required to get only text, so you should use the following code:

```
imdb_khmer = list(pd.read_excel("KHMER.xlsx").text)
```

7. Now that you have text for both languages and the labels, you can split the train and test validations, as follows:

```
from sklearn.model_selection import train_test_split
train_x, test_x, train_y, test_y, khmer_train, khmer_test
= train_test_split(imdb_x, labels, imdb_khmer, test_size
= 0.2, random_state = 1)
```

8. The next step is to provide the representation of these sentences using the XLM-R cross-lingual model. First, you should load the model, as follows:

```
from sentence_transformers import SentenceTransformer
model = SentenceTransformer("stsb-xlm-r-multilingual")
```

9. And now, you can get the representations, like this:

```
encoded_train = model.encode(train_x)
encoded_test = model.encode(test_x)
encoded_khmer_test = model.encode(khmer_test)
```

10. But you should not forget to convert the labels to numpy format because TensorFlow and Keras only deal with numpy arrays when using the `fit` function of the Keras models. Here's how to do it:

```
import numpy as np
train_y = np.array(train_y)
test_y = np.array(test_y)
```

11. Now that everything is ready, let's make a very simple model for classifying the representations, as follows:

```
import tensorflow as tf
input_ = tf.keras.layers.Input((768,))
classification = tf.keras.layers.Dense(
                        1,
                        activation="sigmoid")(input_)
classification_model = \
            tf.keras.Model(input_, classification)
classification_model.compile(
```

```
        loss=tf.keras.losses.BinaryCrossentropy(),
        optimizer="Adam",
        metrics=["accuracy", "Precision", "Recall"])
```

12. You can fit your model using the following function:

```
classification_model.fit(
                x = encoded_train,
                y = train_y,
        validation_data=(encoded_test, test_y),
                epochs = 10)
```

13. And the results for 20 epochs of training are shown, as follows:

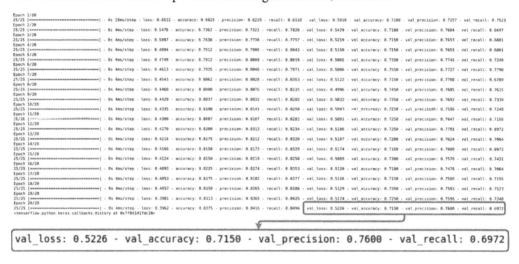

Figure 9.14 – Training results on the English version of the IMDb dataset

14. As you have seen, we used an English test set to see the model performance across epochs, and it is reported as follows in the final epoch:

```
val_loss: 0.5226
val_accuracy: 0.7150
val_precision: 0.7600
val_recall: 0.6972
```

15. Now we have trained our model and tested it on English, let's test it on the Khmer test set, as our model never saw any of the samples either in English or in Khmer. Here's the code to accomplish this:

```
classification_model.evaluate(x = encoded_khmer_test,
                                 y = test_y)
```

Here are the results:

```
loss: 0.5949
accuracy: 0.7250
precision: 0.7014
recall: 0.8623
```

So far, you have learned how it is possible to leverage the capabilities of cross-lingual models in low-resource languages. It makes a huge impact and difference when you can use such a capability in cases where there are very few samples or no samples to train the model on. In the next section, you will learn how it is possible to use zero-shot learning where there are no samples available, even for high-resource languages such as English.

Cross-lingual zero-shot learning

In previous sections, you learned how to perform zero-shot text classification using monolingual models. Using XLM-R for multilingual and cross-lingual zero-shot classification is identical to the approach and code used previously, so we will use **mT5** here.

mT5, which is a massively multilingual pre-trained language model, is based on the encoder-decoder architecture of Transformers and is also identical to **T5**. T5 is pre-trained on English and mT5 is trained on 101 languages from **Multilingual Common Crawl (mC4)**.

The fine-tuned version of mT5 on the XNLI dataset is available from the HuggingFace repository (https://huggingface.co/alan-turing-institute/mt5-large-finetuned-mnli-xtreme-xnli).

The T5 model and its variant, mT5, is a completely text-to-text model, which means it will produce text for any task it is given, even if the task is classification or NLI. So, in the case of inferring this model, extra steps are required. We'll take the following steps:

1. The first step is to load the model and the tokenizer, as follows:

```
from torch.nn.functional import softmax
from transformers import\
```

```
        MT5ForConditionalGeneration, MT5Tokenizer
model_name = "alan-turing-institute/mt5-large-finetuned-
mnli-xtreme-xnli"
tokenizer = MT5Tokenizer.from_pretrained(model_name)
model = MT5ForConditionalGeneration\
        .from_pretrained(model_name)
```

2. In the next step, let's provide samples to be used in zero-shot classification—a sentence and labels, as follows:

```
sequence_to_classify = \
    "Wen werden Sie bei der nächsten Wahl wählen? "
candidate_labels = ["spor", "ekonomi", "politika"]
hypothesis_template = "Dieses Beispiel ist {}."
```

As you see, the sequence itself is in German (`"Who will you vote for in the next election?"`) but the labels are written in Turkish (`"spor"`, `"ekonomi"`, `"politika"`). The hypothesis_template says: `"this example is ..."` in German.

3. The next step is to set the label **identifiers** (IDs) of the entailment, CONTRADICTS, and NEUTRAL, which will be used later in inferring the generated results. Here's the code you'll need to do this:

```
ENTAILS_LABEL = "_0"
NEUTRAL_LABEL = "_1"
CONTRADICTS_LABEL = "_2"
label_inds = tokenizer.convert_tokens_to_ids([
                        ENTAILS_LABEL,
                        NEUTRAL_LABEL,
                        CONTRADICTS_LABEL])
```

4. As you'll recall, the T5 model uses prefixes to know the task that it is supposed to perform. The following function provides the XNLI prefix, along with the premise and hypothesis in the proper format:

```
def process_nli(premise, hypothesis):
    return f'xnli: premise: {premise} hypothesis:
{hypothesis}'
```

5. In the next step, for each label, a sentence will be generated, as illustrated in the following code snippet:

```
pairs =[(sequence_to_classify,\
        hypothesis_template.format(label)) for label in
        candidate_labels]
seqs = [process_nli(premise=premise,
                    hypothesis=hypothesis)
                    for premise, hypothesis in pairs]
```

6. You can see the resulting sequences by printing them, as follows:

```
print(seqs)
['xnli: premise: Wen werden Sie bei der nächsten Wahl
wählen?  hypothesis: Dieses Beispiel ist spor.',
 'xnli: premise: Wen werden Sie bei der nächsten Wahl
wählen?  hypothesis: Dieses Beispiel ist ekonomi.',
 'xnli: premise: Wen werden Sie bei der nächsten Wahl
wählen?  hypothesis: Dieses Beispiel ist politika.']
```

These sequences simply say that the task is XNLI-coded by `xnli :`; the premise sentence is `"Who will you vote for in the next election?"` (in German) and the hypothesis is `"this example is politics"`, `"this example is a sport"`, or `"this example is economy"`.

7. In the next step, you can tokenize the sequences and give them to the model to generate the text according to it, as follows:

```
inputs = tokenizer.batch_encode_plus(seqs,
        return_tensors="pt", padding=True)
out = model.generate(**inputs, output_scores=True,
        return_dict_in_generate=True,num_beams=1)
```

8. The generated text actually gives scores for each token in the vocabulary, and what we are looking for is the entailment, contradiction, and neutral scores. You can get their score using their token IDs, as follows:

```
scores = out.scores[0]
scores = scores[:, label_inds]
```

9. You can see these scores by printing them, like this:

    ```
    >>> print(scores)
    tensor([[-0.9851,  2.2550, -0.0783],
            [-5.1690, -0.7202, -2.5855],
            [ 2.7442,  3.6727,  0.7169]])
    ```

10. The neutral score is not required for our purpose, and we only need contradiction compared to entailment. So, you can use the following code to get only these scores:

    ```
    entailment_ind = 0
    contradiction_ind = 2
    entail_vs_contra_scores = scores[:, [entailment_ind,
    contradiction_ind]]
    ```

11. Now that you have these scores for each sequence of the samples, you can apply a `softmax` layer on it to get the probabilities, as follows:

    ```
    entail_vs_contra_probas = softmax(entail_vs_contra_
    scores, dim=1)
    ```

12. To see these probabilities, you can use `print`, like this:

    ```
    >>> print(entail_vs_contra_probas)
    tensor([[0.2877, 0.7123],
            [0.0702, 0.9298],
            [0.8836, 0.1164]])
    ```

13. Now, you can compare the entailment probability of these three samples by selecting them and applying a `softmax` layer over them, as follows:

    ```
    entail_scores = scores[:, entailment_ind]
    entail_probas = softmax(entail_scores, dim=0)
    ```

14. And to see the values, use `print`, as follows:

    ```
    >>> print(entail_probas)
    tensor([2.3438e-02, 3.5716e-04, 9.7620e-01])
    ```

15. The result means the highest probability belongs to the third sequence. In order to see it in a better shape, use the following code:

```
>>> print(dict(zip(candidate_labels, entail_probas.
tolist())))
{'ekonomi': 0.0003571564157027751,
 'politika': 0.9762046933174133,
 'spor': 0.023438096046447754}
```

The whole process can be summarized as follows: each label is given to the model with the premise, and the model generates scores for each token in the vocabulary. We use these scores to find out how much the entailment token scores over the contradiction.

Fundamental limitations of multilingual models

Although the multilingual and cross-lingual models are promising and will affect the direction of NLP work, they still have some limitations. Many recent works addressed these limitations. Currently, the mBERT model slightly underperforms in many tasks compared with its monolingual counterparts and may not be a potential substitute for a well-trained monolingual model, which is why monolingual models are still widely used.

Studies in the field indicate that multilingual models suffer from the so-called *curse of multilingualism* as they seek to appropriately represent all languages. Adding new languages to a multilingual model improves its performance, up to a certain point. However, it is also seen that adding it after this point degrades performance, which may be due to shared vocabulary. Compared to monolingual models, multilingual models are significantly more limited in terms of the parameter budget. They need to allocate their vocabulary to each one of more than 100 languages.

The existing performance differences between mono- and multilingual models can be attributed to the capability of the designated tokenizer. The study *How Good is Your Tokenizer? On the Monolingual Performance of Multilingual Language Models* (2021) by Rust et al. (`https://arxiv.org/abs/2012.15613`) showed that when a dedicated language-specific tokenizer rather than a general-purpose one (a shared multilingual tokenizer) is attached to a multilingual model, it improves the performance for that language.

Some other findings indicate that it is not currently possible to represent all the world's languages in a single model due to an imbalance in resource distribution of different languages. As a solution, low-resource languages can be oversampled, while high-resource languages can be undersampled. Another observation is that knowledge transfer between two languages can be more efficient if those languages are close. If they are distant languages, this transfer may have little effect. This observation may explain why we got worse results for Afrikaans and Tamil languages in the previous cross-lingual sentence-pair experiment part.

However, there is a lot of work on this subject, and these limitations may be overcome at any time. As of writing this article, the team of XML-R recently proposed two new models—namely, XLM-R XL and XLM-R XXL—that outperform the original XLM-R model by 1.8% and 2.4% average accuracies respectively on XNLI.

Fine-tuning the performance of multilingual models

Now, let's check whether the fine-tuned performance of the multilingual models is actually worse than the monolingual models or not. As an example, let's recall the example of Turkish text classification with seven classes in *Chapter 5*, *Fine-Tuning Language Models for Text Classification*. In that experiment, we fine-tuned a Turkish-specific monolingual model and achieved a good result. We will repeat the same experiment, keeping everything as-is but replacing the Turkish monolingual model with the mBERT and XLM-R models, respectively. Here's how we'll do this:

1. Let's recall the codes in that example again. We had fine-tuned the `"dbmdz/bert-base-turkish-uncased"` model, as follows:

    ```
    from transformers import BertTokenizerFast
    tokenizer = BertTokenizerFast.from_pretrained(
                    "dbmdz/bert-base-turkish-uncased")
    from transformers import BertForSequenceClassification
    model = \ BertForSequenceClassification.from_
    pretrained("dbmdz/bert-base-turkish-uncased",num_
    labels=NUM_LABELS,
                        id2label=id2label,
                        label2id=label2id)
    ```

    With the monolingual model, we got the following performance values:

[77/77 01:24]

	eval_loss	eval_Accuracy	eval_F1	eval_Precision	eval_Recall
train	0.091844	0.975510	0.97546	0.975942	0.975535
val	0.280120	0.924898	0.92381	0.924427	0.924510
test	0.280038	0.926531	0.92542	0.927410	0.925425

Figure 9.15 – Monolingual text classification performance (from Chapter 5, Fine-Tuning Language Models for Text Classification)

2. To fine-tune with mBERT, we need to only replace the preceding model instantiation lines. Now, we will use the `"bert-base-multilingual-uncased"` multilingual model. We instantiate it like this:

```
from transformers import \ BertForSequenceClassification,
AutoTokenizer
tokenizer = AutoTokenizer.from_pretrained(
                "bert-base-multilingual-uncased")
model = BertForSequenceClassification.from_pretrained(
                "bert-base-multilingual-uncased",
                num_labels=NUM_LABELS,
                id2label=id2label,
                label2id=label2id)
```

3. There is not much difference in coding. When we run the experiment keeping all other parameters and settings the same, we get the following performance values:

[77/77 01:26]

	eval_loss	eval_Accuracy	eval_F1	eval_Precision	eval_Recall
train	0.093405	0.978367	0.978373	0.978547	0.978291
val	0.325458	0.911837	0.911586	0.911678	0.911592
test	0.372160	0.904490	0.903152	0.902647	0.904335

Figure 9.16 – mBERT fine-tuned performance

Hmm! The multilingual model underperforms compared with its monolingual counterpart roughly by 2.2% on all metrics.

4. Let's fine-tune the `"xlm-roberta-base"` XLM-R model for the same problem. We'll execute the XLM-R model initialization code, as follows:

```
from transformers import AutoTokenizer,
XLMRobertaForSequenceClassification
tokenizer = AutoTokenizer.from_pretrained(
                              "xlm-roberta-base")
model = XLMRobertaForSequenceClassification\
               .from_pretrained("xlm-roberta-base",
               num_labels=NUM_LABELS,
               id2label=id2label,label2id=label2id)
```

5. Again, we keep all other settings exactly the same. We get the following performance values with the XML-R model:

[77/77 01:26]

	eval_loss	eval_Accuracy	eval_F1	eval_Precision	eval_Recall
train	0.122369	0.968571	0.968665	0.968830	0.968862
val	0.339011	0.912653	0.912454	0.913331	0.912042
test	0.334882	0.915918	0.915662	0.918334	0.914893

Figure 9.17 – XLM-R fine-tuned performance

Not bad! The XLM model did give comparable results. The obtained results are quite close to the monolingual model, with a roughly 1.0% difference. Therefore, although monolingual results can be better than multilingual models in certain tasks, we can achieve promising results with multilingual models. Think of it this way: we may not want to train a whole monolingual model for a 1% performance that lasts 10 days and more. Such small performance differences may be negligible for us.

Summary

In this chapter, you learned about multilingual and cross-lingual language model pre-training and the difference between monolingual and multilingual pre-training. CLM and TLM were also covered, and you gained knowledge about them. You learned how it is possible to use cross-lingual models on various use cases, such as semantic search, plagiarism, and zero-shot text classification. You also learned how it is possible to train on a dataset from a language and test on a completely different language using cross-lingual models. Fine-tuning the performance of multilingual models was evaluated, and we concluded that some multilingual models can be a substitute for monolingual models, remarkably keeping performance loss to a minimum.

In the next chapter, you will learn how to deploy transformer models for real problems and train them for production at an industrial scale.

References

- *Conneau, A., Lample, G., Rinott, R., Williams, A., Bowman, S. R., Schwenk, H. and Stoyanov, V. (2018). XNLI: Evaluating cross-lingual sentence representations. arXiv preprint arXiv:1809.05053.*

- *Xue, L., Constant, N., Roberts, A., Kale, M., Al-Rfou, R., Siddhant, A. and Raffel, C. (2020). mT5: A massively multilingual pre-trained text-to-text transformer. arXiv preprint arXiv:2010.11934.*

- *Lample, G. and Conneau, A. (2019). Cross-lingual language model pretraining. arXiv preprint arXiv:1901.07291.*

- *Conneau, A., Khandelwal, K., Goyal, N., Chaudhary, V., Wenzek, G., Guzmán, F. and Stoyanov, V. (2019). Unsupervised cross-lingual representation learning at scale. arXiv preprint arXiv:1911.02116.*

- *Feng, F., Yang, Y., Cer, D., Arivazhagan, N. and Wang, W. (2020). Language-agnostic bert sentence embedding. arXiv preprint arXiv:2007.01852.*

- *Rust, P., Pfeiffer, J., Vulić, I., Ruder, S. and Gurevych, I. (2020). How Good is Your Tokenizer? On the Monolingual Performance of Multilingual Language Models. arXiv preprint arXiv:2012.15613.*

- *Goyal, N., Du, J., Ott, M., Anantharaman, G. and Conneau, A. (2021). Larger-Scale Transformers for Multilingual Masked Language Modeling. arXiv preprint arXiv:2105.00572.*

10
Serving Transformer Models

So far, we've explored many aspects surrounding Transformers, and you've learned how to train and use a Transformer model from scratch. You also learned how to fine-tune them for many tasks. However, we still don't know how to serve these models in production. Like any other real-life and modern solution, **Natural Language Processing** (**NLP**)-based solutions must be able to be served in a production environment. However, metrics such as response time must be taken into consideration while developing such solutions.

This chapter will explain how to serve a Transformer-based NLP solution in environments where CPU/GPU is available. **TensorFlow Extended** (**TFX**) for machine learning deployment as a solution will be described here. Also, other solutions for serving Transformers as APIs such as FastAPI will be illustrated. You will also learn about the basics of Docker, as well as how to dockerize your service and make it deployable. Lastly, you will learn how to perform speed and load tests on Transformer-based solutions using Locust.

We will cover the following topics in this chapter:

- fastAPI Transformer model serving
- Dockerizing APIs
- Faster Transformer model serving using TFX
- Load testing using Locust

Technical requirements

We will be using Jupyter Notebook, Python, and Dockerfile to run our coding exercises, which will require Python 3.6.0. The following packages need to be installed:

- TensorFlow
- PyTorch
- Transformer >=4.00
- fastAPI
- Docker
- Locust

Now, let's get started!

All the notebooks for the coding exercises in this chapter will be available at the following GitHub link: `https://github.com/PacktPublishing/Mastering-Transformers/tree/main/CH10`.

Check out the following link to see the Code in Action video: `https://bit.ly/375TOPO`

fastAPI Transformer model serving

There are many web frameworks we can use for serving. Sanic, Flask, and fastAPI are just some examples. However, fastAPI has recently gained so much attention because of its speed and reliability. In this section, we will use fastAPI and learn how to build a service according to its documentation. We will also use `pydantic` to define our data classes. Let's begin!

1. Before we start, we must install `pydantic` and fastAPI:

   ```
   $ pip install pydantic
   $ pip install fastapi
   ```

2. The next step is to make the data model for decorating the input of the API using `pydantic`. But before forming the data model, we must know what our model is and identify its input.

   We are going to use a **Question Answering (QA)** model for this. As you know from *Chapter 6, Fine-Tuning Language Models for Token Classification*, the input is in the form of a question and a context.

3. By using the following data model, you can make the QA data model:

   ```
   from pydantic import BaseModel
   class QADataModel(BaseModel):
       question: str
       context: str
   ```

4. We must load the model once and not load it for each request; instead, we will preload it once and reuse it. Because the endpoint function is called each time we send a request to the server, this will result in the model being loaded each time:

   ```
   from transformers import pipeline
   model_name = 'distilbert-base-cased-distilled-squad'
   model = pipeline(model=model_name, tokenizer=model_name,
                               task='question-answering')
   ```

5. The next step is to make a fastAPI instance for moderating the application:

   ```
   from fastapi import FastAPI
   app = FastAPI()
   ```

6. Afterward, you must make a fastAPI endpoint using the following code:

```
@app.post("/question_answering")
async def qa(input_data: QADataModel):
    result = model(question = input_data.question,
context=input_data.context)
    return {"result": result["answer"]}
```

7. It is important to use `async` for the function to make this function run in asynchronous mode; this will be parallelized for requests. You can also use the `workers` parameter to increase the number of workers for the API, as well as making it answer different and independent API calls at once.

8. Using `uvicorn`, you can run your application and serve it as an API. **Uvicorn** is a lightning-fast server implementation for Python-based APIs that makes them run as fast as possible. Use the following code for this:

```
if __name__ == '__main__':
        uvicorn.run('main:app', workers=1)
```

9. It is important to remember that the preceding code must be saved in a `.py` file (`main.py`, for example). You can run it by using the following command:

```
$ python main.py
```

As a result, you will see the following output in your terminal:

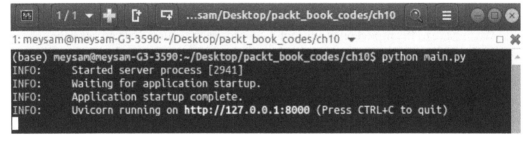

Figure 10.1 – fastAPI in action

10. The next step is to use and test it. There are many tools we can use for this but Postman is one of the best. Before we learn how to use Postman, use the following code:

```
$ curl --location --request POST 'http://127.0.0.1:8000/
question_answering' \
--header 'Content-Type: application/json' \
--data-raw '{
```

```
    "question":"What is extractive question answering?",
    "context":"Extractive Question Answering is the task
of extracting an answer from a text given a question.
An example of a question answering dataset is the SQuAD
dataset, which is entirely based on that task. If you
would like to fine-tune a model on a SQuAD task, you may
leverage the `run_squad.py`."
}'
```

As a result, you will get the following output:

```
{"answer":"the task of extracting an answer from a text
given a question"}
```

Curl is a useful tool but not as handy as Postman. Postman comes with a GUI and is easier to use compared to curl, which is a CLI tool. To use Postman, install it from `https://www.postman.com/downloads/`.

11. After installing Postman, you can easily use it, as shown in the following screenshot:

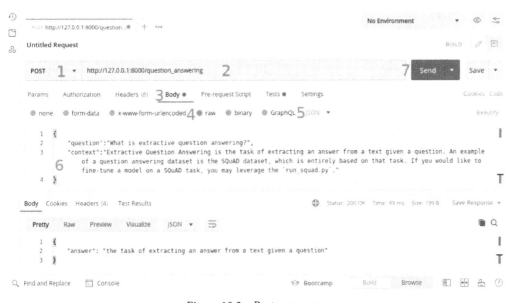

Figure 10.2 – Postman usage

12. Each step for setting up Postman for your service is numbered in the preceding screenshot. Let's take a look at them:

1. Select **POST** as your method.

2. Enter your full endpoint URL.

3. Select **Body**.

4. Set **Body** to **raw**.

5. Select the **JSON** data type.

6. Enter your input data in JSON format.

7. Click **Send**.

You will see the result in the bottom section of Postman.

In the next section, you will learn how to dockerize your fastAPI-based API. It is essential to learn Docker basics to make your APIs packageable and easier for deployment.

Dockerizing APIs

To save time during production and ease the deployment process, it is essential to use Docker. It is very important to isolate your service and application. Also, note that the same code can be run anywhere, regardless of the underlying OS. To achieve this, Docker provides great functionality and packaging. Before using it, you must install it using the steps recommended in the Docker documentation (https://docs.docker.com/get-docker/):

1. First, put the main.py file in the app directory.

2. Next, you must eliminate the last part from your code by specifying the following:

```
if __name__ == '__main__':
    uvicorn.run('main:app', workers=1)
```

3. The next step is to make a Dockerfile for your fastAPI; you made this previously. To do so, you must create a Dockerfile that contains the following content:

```
FROM python:3.7
RUN pip install torch
RUN pip install fastapi uvicorn transformers
EXPOSE 80
COPY ./app /app
CMD ["uvicorn", "app.main:app", "--host", "0.0.0.0",
"--port", "8000"]
```

4. Afterward, you can build your Docker container:

```
$ docker build -t qaapi .
And easily start it:
$ docker run -p 8000:8000 qaapi
```

As a result, you can now access your API using port 8000. However, you can still use Postman, as described in the previous section, *fastAPI Transformer model serving*.

So far, you have learned how to make your own API based on a Transformer model and serve it using fastAPI. You then learned how to dockerize it. It is important to know that there are many options and setups you must learn about regarding Docker; we only covered the basics of Docker here.

In the next section, you will learn how to improve your model serving using TFX.

Faster Transformer model serving using TFX

TFX provides a faster and more efficient way to serve deep learning-based models. But it has some important key points you must understand before you use it. The model must be a saved model type from TensorFlow so that it can be used by TFX Docker or the CLI. Let's take a look:

1. You can perform TFX model serving by using a saved model format from TensorFlow. For more information about TensorFlow saved models, you can read the official documentation at `https://www.tensorflow.org/guide/saved_model`. To make a saved model from Transformers, you can simply use the following code:

```
from transformers import TFBertForSequenceClassification
model = \ TFBertForSequenceClassification.from_
pretrained("nateraw/bert-base-uncased-imdb", from_
pt=True)
model.save_pretrained("tfx_model", saved_model=True)
```

2. Before we understand how to use it to serve Transformers, it is required to pull the Docker image for TFX:

```
$ docker pull tensorflow/serving
```

3. This will pull the Docker container of the TFX being served. The next step is to run the Docker container and copy the saved model into it:

```
$ docker run -d --name serving_base tensorflow/serving
```

4. You can copy the saved file into the Docker container using the following code:

```
$ docker cp tfx_model/saved_model tfx:/models/bert
```

5. This will copy the saved model files into the container. However, you must commit the changes:

```
$ docker commit --change "ENV MODEL_NAME bert" tfx my_
bert_model
```

6. Now that everything is ready, you can kill the Docker container:

```
$ docker kill tfx
```

This will stop the container from running.

Now that the model is ready and can be served by the TFX Docker, you can simply use it with another service. The reason we need another service to call TFX is that the Transformer-based models have a special input format provided by tokenizers.

7. To do so, you must make a fastAPI service that will model the API that was served by the TensorFlow serving container. Before you code your service, you should start the Docker container by giving it parameters to run the BERT-based model. This will help you fix bugs in case there are any errors:

```
$ docker run -p 8501:8501 -p 8500:8500 --name bert my_
bert_model
```

8. The following code contains the content of the main.py file:

```
import uvicorn
from fastapi import FastAPI
from pydantic import BaseModel
from transformers import BertTokenizerFast, BertConfig
import requests
import json
import numpy as np
tokenizer =\
  BertTokenizerFast.from_pretrained("nateraw/bert-base-
uncased-imdb")
config = BertConfig.from_pretrained("nateraw/bert-base-
uncased-imdb")
class DataModel(BaseModel):
    text: str
```

```
app = FastAPI()
@app.post("/sentiment")
async def sentiment_analysis(input_data: DataModel):
    print(input_data.text)
    tokenized_sentence = [dict(tokenizer(input_data.
text))]
    data_send = {"instances": tokenized_sentence}
    response = \    requests.post("http://localhost:8501/
v1/models/bert:predict", data=json.dumps(data_send))
    result = np.abs(json.loads(response.text)
["predictions"][0])
    return {"sentiment": config.id2label[np.
argmax(result)]}
if __name__ == '__main__':
    uvicorn.run('main:app', workers=1)
```

9. We have loaded the `config` file because the labels are stored in it, and we need them to return it in the result. You can simply run this file using `python`:

```
$ python main.py
```

Now, your service is up and ready to use. You can access it using Postman, as shown in the following screenshot:

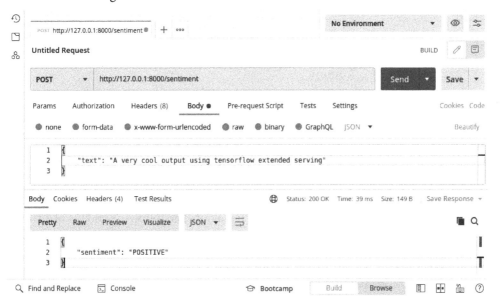

Figure 10.3 – Postman output of a TFX-based service

The overall architecture of the new service within TFX Docker is shown in the following diagram:

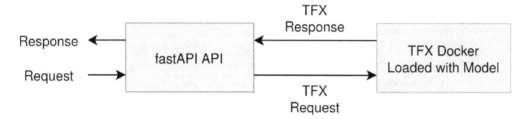

Figure 10.4 – TFX-based service architecture

So far, you have learned how to serve a model using TFX. However, you need to learn how to load test your service using Locust. It is important to know the limits of your service and when to optimize it by using quantization or pruning. In the next section, we will describe how to test model performance under heavy load using Locust.

Load testing using Locust

There are many applications we can use to load test services. Most of these applications and libraries provide useful information about the response time and delay of the service. They also provide information about the failure rate. Locust is one of the best tools for this purpose. We will use it to load test three methods for serving a Transformer-based model: using fastAPI only, using dockerized fastAPI, and TFX-based serving using fastAPI. Let's get started:

1. First, we must install Locust:

```
$ pip install locust
```

This command will install Locust. The next step is to make all the services serving an identical task use the same model. Fixing two of the most important parameters of this test will ensure that all the services have been designed identically to serve a single purpose. Using the same model will help us freeze anything else and focus on the deployment performance of the methods.

2. Once everything is ready, you can start load testing your APIs. You must prepare a locustfile to define your user and its behavior. The following code is of a simple locustfile:

```
from locust import HttpUser, task
from random import choice
from string import ascii_uppercase
```

```
class User(HttpUser):
    @task
    def predict(self):
        payload = {"text": ''.join(choice(ascii_
uppercase) for i in range(20))}
        self.client.post("/sentiment", json=payload)
```

By using `HttpUser` and creating the `User` class that's inherited from it, we can define an `HttpUser` class. The `@task` decorator is essential for defining the task that the user must perform after spawning. The `predict` function is the actual task that the user will perform repeatedly after spawning. It will generate a random string that's `20` in length and send it to your API.

3. To start the test, you must start your service. Once you've started your service, run the following code to start the Locust load test:

```
$ locust -f locust_file.py
```

Locust will start with the settings you provided in your `locustfile`. You will see the following in your Terminal:

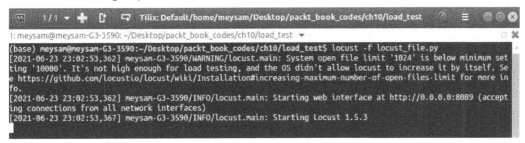

Figure 10.5 – Terminal after starting a Locust load test

As you can see, you can open the URL where the load web interface is located; that is, `http://0.0.0.0:8089`.

4. After opening the URL, you will see an interface, as shown in the following screenshot:

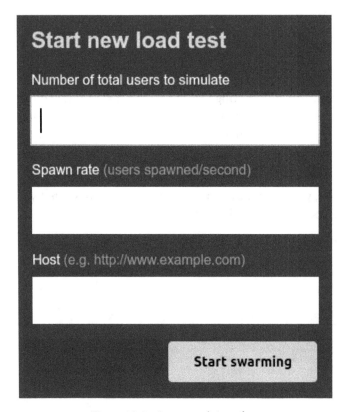

Figure 10.6 – Locust web interface

5. We are going to set **Number of total users to simulate** to **10**, **Spawn rate** to **1**, and **Host** to **http://127.0.0.1:8000**, which is where our service is running. After setting these parameters, click **Start swarming**.

6. At this point, the UI will change, and the test will begin. To stop the test at any time, click the **Stop** button.

7. You can also click the **Charts** tab to see a visualization of the results:

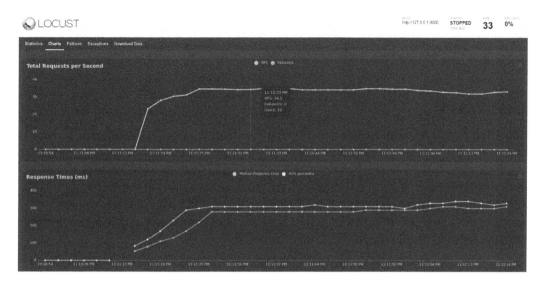

Figure 10.7 – Locust test results from the Charts tab

8. Now that the test is ready for the API, let's test all three versions and compare the results to see which one performs better. Remember that the services must be tested independently on the machine where you want to serve them. In other words, you must run one service at a time and test that, close the service, run the other one and test it, and so on.

The results are shown in the following table:

	TFX-based FastAPI	FastAPI	Dockerized FastAPI
RPS	38.5	33	34
Average RT (ms)	237	275	270

Table 1 – Comparing the results of different implementations

In the preceding table, **Requests Per Second (RPS)** means the number of requests per second that the API answers, while the **Average Response Time (RT)** means the milliseconds that service takes to respond to a given call. These results shows that the TFX-based fastAPI is the fastest. It has a higher RPS and a lower average RT. All these tests were performed on a machine with an Intel(R) Core(TM) i7-9750H CPU with 32 GB RAM, and GPU disabled.

In this section, you learned how to test your API and measure its performance in terms of important parameters such as RPS and RT. However, there are many other stress tests a real-world API can perform, such as increasing the number of users to make them behave like real users. To perform such tests and report their results in a more realistic way, it is important to read Locust's documentation and learn how to perform more advanced tests.

Summary

In this chapter, you learned the basics of serving Transformer models using fastAPI. You also learned how to serve models in a more advanced and efficient way, such as by using TFX. You then studied the basics of load testing and creating users. Making these users spawn in groups or one by one, and then reporting the results of stress testing, was another major topic of this chapter. After that, you studied the basics of Docker and how to package your application in the form of a Docker container. Finally, you learned how to serve Transformer-based models.

In the next chapter, you will learn about Transformer deconstruction, the model view, and monitoring training using various tools and techniques.

References

- Locust documentation: `https://docs.locust.io`
- TFX documentation: `https://www.tensorflow.org/tfx/guide`
- FastAPI documentation: `https://fastapi.tiangolo.com`
- Docker documentation: `https://docs.docker.com`
- HuggingFace TFX serving: `https://huggingface.co/blog/tf-serving`

11
Attention Visualization and Experiment Tracking

In this chapter, we will cover two different technical concepts, **attention visualization** and **experiment tracking**, and we will practice them through sophisticated tools such as **exBERT** and **BertViz**. These tools provide important functions for interpretability and explainability. First, we will discuss how to visualize the inner parts of attention by utilizing the tools. It is important to interpret the learned representations and to understand the information encoded by self-attention heads in the Transformer. We will see that certain heads correspond to a certain aspect of syntax or semantics. Secondly, we will learn how to track experiments by logging and then monitoring by using **TensorBoard** and **Weights & Biases** (**W&B**). These tools enable us to efficiently host and track experimental results such as loss or other metrics, which helps us to optimize model training. You will learn how to use exBERT and BertViz to see the inner parts of their own models and will be able to utilize both TensorBoard and W&B to monitor and optimize their models by the end of the chapter.

We will cover the following topics in this chapter:

- Interpreting attention heads
- Tracking model metrics

Technical requirements

The code for this chapter is found at `https://github.com/PacktPublishing/Mastering-Transformers/tree/main/CH11`, which is the GitHub repository for this book. We will be using Jupyter Notebook to run our coding exercises that require Python 3.6.0 or above, and the following packages will need to be installed:

- `tensorflow`
- `pytorch`
- `Transformers >=4.00`
- `tensorboard`
- `wandb`
- `bertviz`
- `ipywidgets`

Check out the following link to see Code in Action Video:

`https://bit.ly/3iM4Y1F`

Interpreting attention heads

As with most **Deep Learning** (**DL**) architectures, both the success of the Transformer models and how they learn have been not fully understood, but we know that the Transformers—remarkably—learn many linguistic features of the language. A significant amount of learned linguistic knowledge is distributed both in the hidden state and in the self-attention heads of the pre-trained model. There have been substantial recent studies published and many tools developed to understand and to better explain the phenomena.

Thanks to some **Natural Language Processing** (**NLP**) community tools, we are able to interpret the information learned by the self-attention heads in a Transformer model. The heads can be interpreted naturally, thanks to the weights between tokens. We will soon see that in further experiments in this section, certain heads correspond to a certain aspect of syntax or semantics. We can also observe surface-level patterns and many other linguistic features.

In this section, we will conduct some experiments using community tools to observe these patterns and features in the attention heads. Recent studies have already revealed many of the features of self-attention. Let's highlight some of them before we get into the experiments. For example, most of the heads attend to delimiter tokens such as **Separator** (**SEP**) and **Classification** (**CLS**), since these tokens are never masked out and bear segment-level information in particular. Another observation is that most heads pay little attention to the current token, but some heads specialize in only attending the next or previous tokens, especially in earlier layers. Here is a list of other patterns found in recent studies that we can easily observe in our experiments:

- Attention heads in the same layer show similar behavior.

- Particular heads correspond to specific aspects of syntax or semantic relations.

- Some heads encode so that the direct objects tend to attend to their verbs, such as *<lesson, take>* or *<car, drive>*.

- In some heads, the noun modifiers attend to their noun (for example, *the hot water*; *the next layer*), or the possessive pronoun attends to the head (for example, *her car*).

- Some heads encode so that passive auxiliary verbs attend to a related verb, such as *Been damaged, was taken.*

- In some heads, coreferent mentions attend to themselves, such as *talks-negotiation, she-her, President-Biden.*

- The lower layers usually have information about word positions.

- Syntactic features are observed earlier in the transformer, while high-level semantic information appears in the upper layers.

- The final layers are the most task-specific and are therefore very effective for downstream tasks.

To observe these patterns, we can use two important tools, **exBERT** and **BertViz**, here. These tools have almost the same functionality. We will start with exBERT.

Visualizing attention heads with exBERT

exBERT is a visualization tool to see the inner parts of Transformers. We will use it to visualize the attention heads of the *BERT-base-cased* model, which is the default model in the exBERT interface. Unless otherwise stated, the model we will use in the following examples is *BERT-base-cased*. This contains 12 layers and 12 self-attention heads in each layer, which makes for 144 self-attention heads.

We will learn how to utilize exBERT step by step, as follows:

1. Let's click on the exBERT link hosted by *Hugging Face*: `https://huggingface.co/exbert`.

2. Enter the sentence **The cat is very sad.** and see the output, as follows:

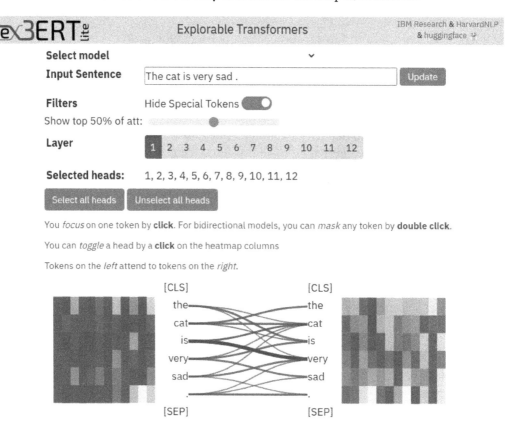

Figure 11.1 – exBERT interface

In the preceding screenshot, the left tokens attend to the right tokens. The thickness of the lines represents the value of the weights. Because CLS and SEP tokens have very frequent and dense connections, we cut off the links associated with them for simplicity. Please see the **Hide Special Tokens** toggle switch. What we see now is the attention mapping at layer 1, where the lines correspond to the sum of the weights on all heads. This is called a **multi-head attention mechanism**, in which 12 heads work in parallel to each other. This mechanism allows us to capture a wider range of relationships than is possible with single-head attention. This is why we see a broadly attending pattern in *Figure 11.1*. We can also observe any specific head by clicking the **Head** column.

If you hover over a token at the left, you will see the specific weights of that token connecting to the right ones. For more detailed information on using the interface, read the paper *exBERT: A Visual Analysis Tool to Explore Learned Representations in Transformer Models, Benjamin Hoover, Hendrik Strobelt, Sebastian Gehrmann, 2019* or watch the video at the following link: `https://exbert.net/`.

3. Now, we will try to support the findings of other researchers addressed in the introductory part of this section. Let's take the *some heads specialize in only attending the next or previous tokens, especially in earlier layers* pattern, and see if there's a head that supports this.

4. We will use **<Layer-No, Head-No>** notation to denote a certain self-attention head for the rest of the chapter, where the indices start at 1 for exBERT and start at 0 for BertViz—for example, *<3,7>* denotes the seventh head at the third layer for exBERT. When you select the **<2,5> (or <4,12> or <6,2 >)** head, you will get the following output, where each token attends to the previous token only:

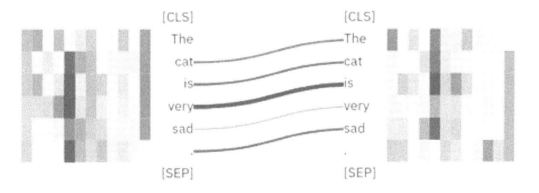

Figure 11.2 – Previous-token attention pattern

5. For the **<2, 12>** and **<3, 4>** heads, you will get a pattern whereby each token attends to the next token, as follows:

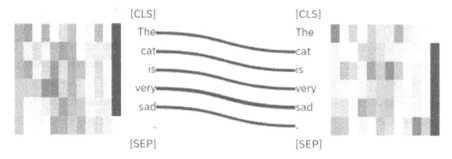

Figure 11.3 – Next-token attention pattern

These heads serve the same functionality for other input sentences—that is, they work independently of the input. You can try different sentences yourself.

We can use an attention head for advanced semantic tasks such as pronoun resolution using a **probing classifier**. First, we will qualitatively check if the internal representation has such a capacity for pronoun resolution (or coreference resolution) or not. Pronoun resolution is considered a challenging semantic relation task since the distance between the pronoun and its antecedent is usually very long.

6. Now, we take the sentence *The cat is very sad. Because it could not find food to eat.* When you check each head, you will notice that the *<9,9>* and *<9,12>* heads encode the pronoun relation. When hovering over **it** at the *<9,9>* head, we get the following output:

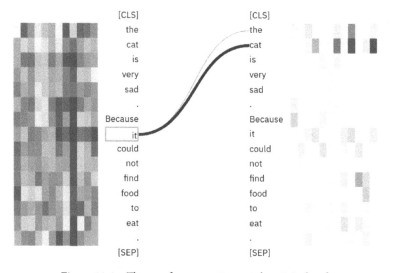

Figure 11.4 – The coreference pattern at the <9,9> head

The *<9,12>* head also works for pronoun relation. Again, on hovering over **it**, we get the following output:

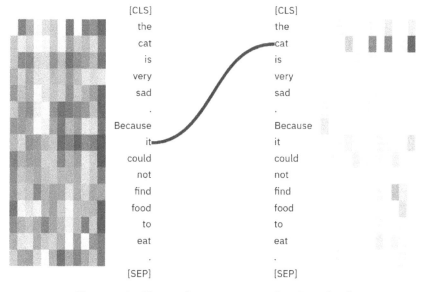

Figure 11.5 – The coreference pattern at the <9,12> head

From the preceding screenshot, we see that the **it** pronoun strongly attends to its antecedent, **cat**. We change the sentence a bit so that the **it** pronoun now refers to the **food** token instead of **cat**, as in **the cat did not eat the food because it was not fresh**. As seen in the following screenshot, which relates to the *<9,9>* head, **it** properly attends to its antecedent **food**, as expected:

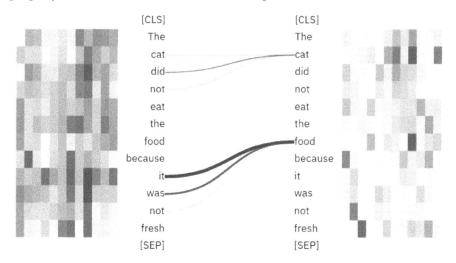

Figure 11.6 – The pattern at the <9,9> head for the second example

7. Let's take another run, where the pronoun refers to **cat**, as in **The cat did not eat the food because it was very angry**. In the <9,9> head, the **it** token mostly attends to the **cat** token, as shown in the following screenshot:

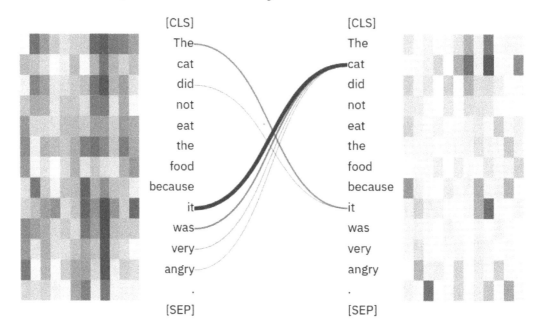

Figure 11.7 – The pattern at the <9,9> head for the second input

8. I think those are enough examples. Now, we will use the exBERT model differently to evaluate the model capacity. Let's restart the exBERT interface, select the last layer (*layer 12*), and keep all heads. Then, enter the sentence **the cat did not eat the food.** and mask out the **food** token. Double-clicking masks the **food** token out, as follows:

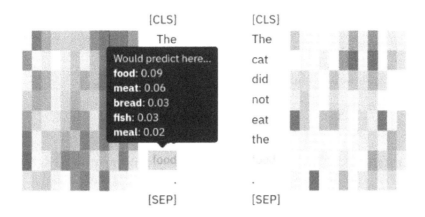

Figure 11.8 – Evaluating the model by masking

When you hover on that masked token, you can see the prediction distribution of the *Bert-base-cased* model, as shown in the preceding screenshot. The first prediction is **food**, which is expected. For more detailed information about the tool, you can use exBERT's web page, at `https://exbert.net/`.

Well done! In the next section, we will work with BertViz and write some Python code to access the attention heads.

Multiscale visualization of attention heads with BertViz

Now, we will write some code to visualize heads with BertViz, which is a tool to visualize attention in the Transformer model, as is exBERT. It was developed by Jesse Vig in 2019 (*A Multiscale Visualization of Attention in the Transformer Model, Jesse Vig, 2019*). It is the extension of the work of the Tensor2Tensor visualization tool (Jones, 2017). We can monitor the inner parts of a model with multiscale qualitative analysis. The advantage of BertViz is that we can work with most Hugging Face-hosted models (such as **Bidirectional Encoder Representations from Transformers** (**BERT**), **Generated Pre-trained Transformer** (**GPT**), and **Cross-lingual Language Model** (**XLM**)) through the Python **Application Programming Interface** (**API**). Therefore, we will be able to work with non-English models as well, or any pre-trained model. We will examine such examples together shortly. You can access BertViz resources and other information from the following GitHub link: `https://github.com/jessevig/bertviz`.

As with exBERT, BertViz visualizes attention heads in a single interface. Additionally, it supports a **bird's eye view** and a low-level **neuron view**, where we observe how individual neurons interact to build attention weights. A useful demonstration video can be found at the following link: `https://vimeo.com/340841955`.

Before starting, we need to install the necessary libraries, as follows:

```
!pip install bertviz ipywidgets
```

We then import the following modules:

```
from bertviz import head_view
from Transformers import BertTokenizer, BertModel
```

BertViz supports three views: a **head view**, a **model view**, and a **neuron view**. Let's examine these views one by one. First of all, though, it is important to point out that we started from 1 to index layers and heads in exBERT. But in BertViz, we start from 0 for indexing, as in Python programming. If I say a *<9,9>* head in exBERT, its BertViz counterpart is *<8,8>*.

Let's start with the head view.

Attention head view

The head view is the BertViz equivalent of what we have experienced so far with exBERT in the previous section. The **attention head view** visualizes the attention patterns based on one or more attention heads in a selected layer:

1. First, we define a `get_bert_attentions()` function to retrieve attentions and tokens for a given model and a given pair of sentences. The function definition is shown in the following code block:

    ```
    def get_bert_attentions(model_path, sentence_a,
    sentence_b):
        model = BertModel.from_pretrained(model_path,
            output_attentions=True)
        tokenizer = BertTokenizer.from_pretrained(model_path)
        inputs = tokenizer.encode_plus(sentence_a,
            sentence_b, return_tensors='pt',
            add_special_tokens=True)
        token_type_ids = inputs['token_type_ids']
        input_ids = inputs['input_ids']
        attention = model(input_ids,
            token_type_ids=token_type_ids)[-1]
        input_id_list = input_ids[0].tolist()
        tokens = tokenizer.convert_ids_to_tokens(input_id_
    ```

```
list)
    return attention, tokens
```

2. In the following code snippet, we load the `bert-base-cased` model and retrieve the tokens and corresponding attentions of the given two sentences. We then call the `head_view()` function at the end to visualize the attentions. Here is the code execution:

```
model_path = 'bert-base-cased'
sentence_a = "The cat is very sad."
sentence_b = "Because it could not find food to eat."
attention, tokens=get_bert_attentions(model_path,
    sentence_a, sentence_b)
head_view(attention, tokens)
```

The code output is an interface, as displayed here:

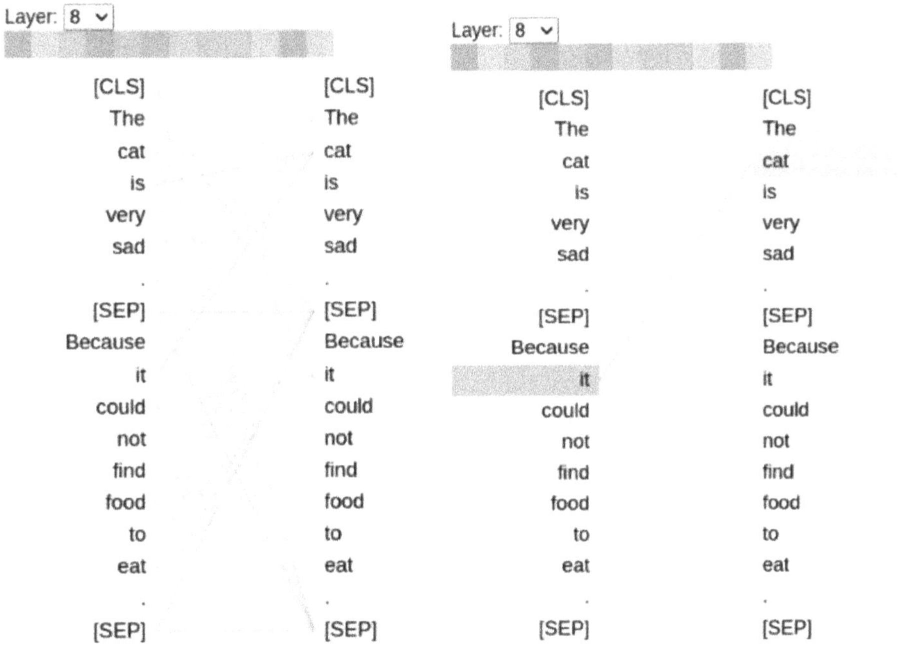

Figure 11.9 – Head-view output of BertViz

The interface on the left of *Figure 11.9* comes first. Hovering over any token on the left will show the attention going from that token. The colored tiles at the top correspond to the attention head. Double-clicking on any of them will select it and discard the rest. The thicker attention lines denote higher attention weights.

Please remember that in the preceding exBERT examples, we observed that the <9,9> head (the equivalent in BertViz is <8, 8>, due to indexing) bears a pronoun-antecedent relationship. We observe the same pattern in *Figure 11.9*, selecting layer 8 and head 8. Then, we see the interface on the right of *Figure 11.9* when we hover on *it*, where *it* strongly attends to the *cat* and *it* tokens. So, can we observe these semantic patterns in other pre-trained language models? Although the heads are not exactly the same in other models, some heads can encode these semantic properties. We also know from recent work that semantic features are mostly encoded in the higher layers.

3. Let's look for a coreference pattern in a Turkish language model. The following code loads a Turkish `bert-base` model and takes a sentence pair. We observe here that the <8,8> head has the same semantic feature in Turkish as in the English model, as follows:

```
model_path = 'dbmdz/bert-base-turkish-cased'
sentence_a = "Kedi çok üzgün."
sentence_b = "Çünkü o her zamanki gibi çok fazla yemek
yedi."
attention, tokens=\
get_bert_attentions(model_path, sentence_a, sentence_b)
head_view(attention, tokens)
```

From the preceding code, `sentence_a` and `sentence_b` mean *The cat is sad* and *Because it ate too much food as usual*, respectively. When hovering over **o** (**it**), **it** attends to **Kedi** (**cat**), as follows:

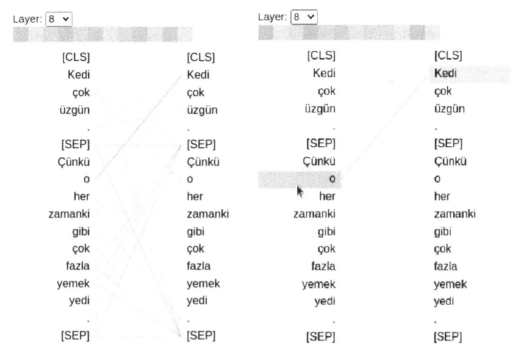

Figure 11.10 – Coreference pattern in the Turkish language model

All other tokens except **o** mostly attend to the SEP delimiter token, which is a dominant behavior pattern in all heads in the BERT architecture.

4. As a final example for the head view, we will interpret another language model and move on to the model view feature. This time, we choose the `bert-base-german-cased` German language model and visualize it for the input—that is, the German equivalent of the same-sentence pair we used for Turkish.

5. The following code loads a German model, consumes a pair of sentences, and visualizes them:

```
model_path = 'bert-base-german-cased'
sentence_a = "Die Katze ist sehr traurig."
sentence_b = "Weil sie zu viel gegessen hat"
attention, tokens=\
get_bert_attentions(model_path, sentence_a, sentence_b)
head_view(attention, tokens)
```

6.   When we examine the heads, we can see the coreference pattern in the 8th layer again, but this time in the 11th head. To select the *<8,11>* head, pick layer 8 from the drop-down menu and double-click on the last head, as follows:

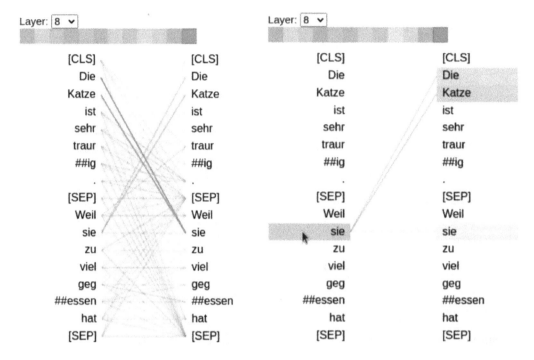

Figure 11.11 – Coreference relation pattern in the German language model

As you see, when hovering over **sie**, you will see strong attentions to the **Die Katze**. While this *<8,11>* head is the strongest one for coreference relations (known as anaphoric relations in computational linguistics literature), this relationship may have spread to many other heads. To observe it, we will have to check all the heads one by one.

On the other hand, BertViz's model view feature gives us a basic bird's-eye view to see all heads at once. Let's take a look at it in the next section.

Model view

Model view allows us to have a bird's-eye view of attentions across all heads and layers. Self-attention heads are shown in tabular form, with rows and columns corresponding to layers and heads, respectively. Each head is visualized in the form of a clickable thumbnail that includes the broad shape of the attention model.

The view can tell us how BERT works and makes it easier to interpret. Many recent studies, such as *A Primer in BERTology: What We Know About How BERT Works, Anna Rogers, Olga Kovaleva, Anna Rumshisky, 2021*, found some clues about the behavior of the layers and came to a consensus. We already listed some of them in the *Interpreting attention heads* section. You can test these facts yourself using BertViz's model view.

Let's view the German language model that we just used, as follows:

1. First, import the following modules:

```
from bertviz import model_view
from Transformers import BertTokenizer, BertModel
```

2. Now, we will use a `show_model_view()` wrapper function developed by Jesse Vig. You can find the original code at the following link: `https://github.com/jessevig/bertviz/blob/master/notebooks/model_view_bert.ipynb`.

3. You can also find the function definition in our book's GitHub link, at `https://github.com/PacktPublishing/Mastering-Transformers/tree/main/CH11`. We are just dropping the function header here:

```
def show_model_view(model, tokenizer, sentence_a,
    sentence_b=None, hide_delimiter_attn=False,
    display_mode="dark"):
    . . .
```

4. Let's load the German model again. If you have already loaded it, you can skip the first five lines. Here is the code you'll need:

```
model_path='bert-base-german-cased'
sentence_a = "Die Katze ist sehr traurig."
sentence_b = "Weil sie zu viel gegessen hat"
model = BertModel.from_pretrained(model_path, output_
attentions=True)
tokenizer = BertTokenizer.from_pretrained(model_path)
show_model_view(model, tokenizer, sentence_a, sentence_b,
    hide_delimiter_attn=False,
    display_mode="light")
```

This is the output:

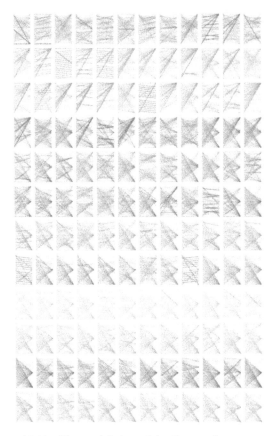

Figure 11.12 – The model view of the German language model

This view helps us easily observe many patterns such as next-token (or previous-token) attention patterns. As we mentioned earlier in the *Interpreting attention heads* section, tokens often tend to attend to delimiters—specifically, CLS delimiters at lower layers and SEP delimiters at upper layers. Because these tokens are not masked out, they can ease the flow of information. In the last layers, we only observe SEP-delimiter-focused attention patterns. It could be speculated that SEP is used to collect segment-level information, which can be used then for inter-sentence tasks such as **Next Sentence Prediction** (**NSP**) or for encoding sentence-level meaning.

On the other hand, we observe that coreference relation patterns are mostly encoded in the *<8,1>*, *<8,11>*, *<10,1>*, *and <10,7>* heads. Again, it can be clearly said that the *<8, 11>* head is the strongest head that encodes the coreference relation in the German model, which we already discussed.

5.   When you click on that thumbnail, you will see the same output, as follows:

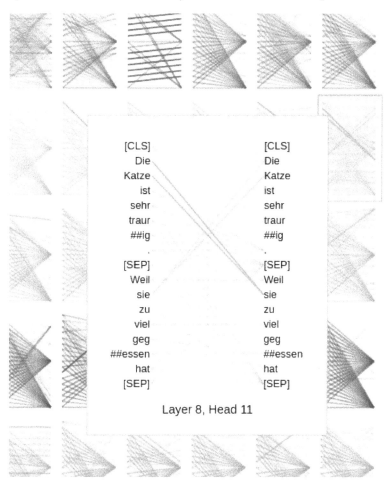

Figure 11.13 – Close-up of the <8,11> head in model view

You can again hover over the tokens and see the mappings.

I think that's enough work for the head view and the model view. Now, let's deconstruct the model with the help of the neuron view and try to understand how these heads calculate weights.

Neuron view

So far, we have visualized computed weights for a given input. The **neuron view** visualizes the neurons and the key vectors in a query and how the weights between tokens are computed based on interactions. We can trace the computation phase between any two tokens.

Again, we will load the German model and visualize the same-sentence pair we just worked with, to be coherent. We execute the following code:

```
from bertviz.Transformers_neuron_view import BertModel,
BertTokenizer
from bertviz.neuron_view import show
model_path='bert-base-german-cased'
sentence_a = "Die Katze ist sehr traurig."
sentence_b = "Weil sie zu viel gegessen hat"
model = BertModel.from_pretrained(model_path, output_
attentions=True)
tokenizer = BertTokenizer.from_pretrained(model_path)
model_type = 'bert'
show(model, model_type, tokenizer, sentence_a, sentence_b,
layer=8, head=11)
```

This is the output:

Figure 11.14 – Neuron view of the coreference relation pattern (head <8,11>)

The view helps us to trace the computation of attention from the **sie** token that we selected on the left to the other tokens on the right. Positive values are blue and negative values are orange. Color intensity represents the magnitude of the numerical value. The query of **sie** is very similar to the keys of **Die** and **Katze**. If you look at the patterns carefully, you will notice how similar these vectors are. Therefore, their dot product goes higher than the other comparison, which establishes strong attention between those tokens. We also trace the dot product and the Softmax function output as we go to the right. When clicking on the other tokens on the left, you can trace other computations as well.

Now, let's select a head-bearing next-token attention pattern for the same input, and trace it. To do so, we select the *<2,6>* head. In this pattern, virtually all the attention is focused on the next word. We click the **sie** token once again, as follows:

Figure 11.15 – Neuron view of next-token attention patterns (the <2,6> head)

Now, the **sie** token is focused on the next token instead of its own antecedent (**Die Katze**). When we carefully look at the query and the candidate keys, the most similar key to the query of **sie** is the next token, **zu**. Likewise, we observe how the dot product and Softmax function are applied in order.

In the next section, we will briefly talk about probing classifiers for interpreting Transformers.

Understanding the inner parts of BERT with probing classifiers

The opacity of what DL learns has led to a number of studies on the interpretation of such models. We attempt to answer the question of which parts of a Transformer model are responsible for certain language features, or which parts of the input lead the model to make a particular decision. To do so, other than visualizing internal representations, we can train a classifier on the representations to predict some external morphological, syntactic, or semantic properties. Hence, we can determine if we associate internal *representations* with external *properties*. The successful training of the model would be quantitative evidence of such an association—that is, the language model has learned information relevant for an external property. This approach is called a **probing-classifier** approach, which is a prominent analysis technique in NLP and other DL studies. An attention-based probing classifier takes an attention map as input and predicts external properties such as coreference relations or head-modifier relations.

As seen in the preceding experiments, we get the self-attention weights for a given input with the get_bert_attention() function. Instead of visualizing these weights, we can directly transfer them to a classification pipeline. So, with supervision, we can determine which head is suitable for which semantic feature—for example, we can figure out which heads are suitable for coreference with labeled data.

Now, let's move on to the model-tracking part, which is crucial for building efficient models.

Tracking model metrics

So far, we have trained language models and simply analyzed the final results. We have not observed the training process or made a comparison of training using different options. In this section, we will briefly discuss how to monitor model training. For this, we will handle how to track the training of the models we developed before in *Chapter 5, Fine-Tuning Language Models for Text Classification*.

There are two important tools developed in this area—one is TensorBoard, and the other is W&B. With the former, we save the training results to a local drive and visualize them at the end of the experiment. With the latter, we are able to monitor the model-training progress live in a cloud platform.

This section will be a short introduction to these tools without going into much detail about them, as this is beyond the scope of this chapter.

Let's start with TensorBoard.

Tracking model training with TensorBoard

TensorBoard is a visualization tool specifically for DL experiments. It has many features such as tracking, training, projecting embeddings to a lower space, and visualizing model graphs. We mostly use it for tracking and visualizing metrics such as loss. Tracking a metric with TensorBoard is so easy for Transformers that adding a couple of lines to model-training code will be enough. Everything is kept almost the same.

Now, we will repeat the **Internet Movie Database (IMDb)** sentiment fine-tuning experiment we did in *Chapter 5*, *Fine-Tuning Language Models for Text Classification*, and will track the metrics. In that chapter, we already trained a sentiment model with an IMDb dataset consisting of a **4 kilo (4K)** training dataset, a 1K validation set, and a 1K test set. Now, we will adapt it to TensorBoard. For more details about TensorBoard, please visit `https://www.tensorflow.org/tensorboard`.

Let's begin:

1. First, we install TensorBoard if it is not already installed, like this:

    ```
    !pip install tensorboard
    ```

2. Keeping the other code lines of IMDb sentiment analysis as-is from *Chapter 5*, *Fine-Tuning Language Models for Text Classification*, we set the training argument as follows:

    ```
    from Transformers import TrainingArguments, Trainer
    training_args = TrainingArguments(
        output_dir='./MyIMDBModel',
        do_train=True,
        do_eval=True,
        num_train_epochs=3,
        per_device_train_batch_size=16,
        per_device_eval_batch_size=32,
        logging_strategy='steps',
        logging_dir='./logs',
        logging_steps=50,
        evaluation_strategy="steps",
        save_strategy="epoch",
        fp16=True,
        load_best_model_at_end=True
    )
    ```

3. In the preceding code snippet, the value of `logging_dir` will soon be passed to TensorBoard as a parameter. As the training dataset size is 4K and the training batch size is 16, we have 250 steps (4K/16) for each epoch, which means 750 steps for three epochs.

4. We set `logging_steps` to 50, which is a sampling interval. As the interval is decreased, more details about where model performance rises or falls are recorded. We'll do another experiment later on, reducing this sampling interval at step 27.

5. Now, at every 50 steps, the model performance is measured in terms of the metrics that we define in `compute_metrics()`. The metrics to be measured are Accuracy, F1, Precision, and Recall. As a result, we will have 15 (750/50) performance measurements to be recorded. When we run `trainer.train()`, this starts the training process and records the logs under the `logging_dir='./logs'` directory.

6. We set `load_best_model_at_end` to `True` so that the pipeline loads whichever checkpoint has the best performance in terms of loss. Once the training is completed, you will notice that the best model is loaded from `checkpoint-250` with a loss score of `0.263`.

7. Now, the only thing we need to do is to call the following code to launch TensorBoard:

```
%reload_ext tensorboard
%tensorboard --logdir logs
```

This is the output:

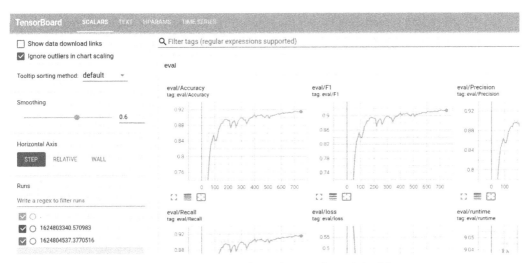

Figure 11.16 – TensorBoard visualization for training history

As you may have noticed, we can trace the metrics that we defined before. The horizontal axis goes from 0 to 750 steps, which is what we calculated before. We will not discuss TensorBoard in detail here. Let's just look at the **eval/loss** chart only. When you click on the maximization icon at the left-hand bottom corner, you will see the following chart:

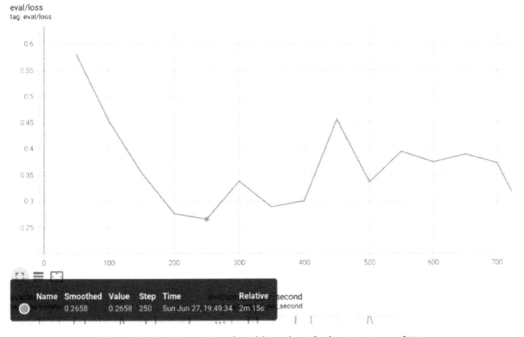

Figure 11.17 – TensorBoard eval/loss chart for logging steps of 50

In the preceding screenshot, we set the smoothing to 0 with the slider control on the left of the TensorBoard dashboard to see scores more precisely and focus on the global minimum. If your experiment has very high volatility, the smoothing feature can work well to see overall trends. It functions as a **Moving Average (MA)**. This chart supports our previous observation, in which the best loss measurement is **0.2658** at step **250**.

8. As `logging_steps` is set to **10**, we get a high resolution, as in the following screenshot. As a result, we will have 75 (750 steps/10 steps) performance measurements to be recorded. When we rerun the entire flow with this resolution, we get the best model at step 220, with a loss score of 0.238, which is better than the previous experiment. The result can be seen in the following screenshot. We naturally observe more fluctuations due to higher resolution:

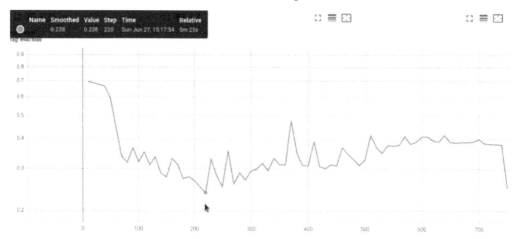

Figure 11.18 – Higher-resolution eval/loss chart for logging steps of 10

We are done with TensorBoard for now. Let's work with W&B!

Tracking model training live with W&B

W&B, unlike TensorBoard, provides a dashboard in a cloud platform, and we can trace and back up all experiments in a single hub. It also allows us to work with a team for development and sharing. The training code is run on our local machine, while the logs are kept in the W&B cloud. Most importantly, we can follow the training process live and share the result immediately with the community or team.

We can enable W&B for our experiments by making very small changes to our existing code:

1. First of all, we need to create an account in `wandb.ai`, then install the Python library, as follows:

    ```
    !pip install wandb
    ```

2. Again, we will take the IMDb sentiment-analysis code and make minor changes to it. First, let's import the library and log in to wandB, as follows:

    ```
    import wandb
    !wandb login
    ```

    wandb requests an API key that you can easily find at the following link: `https://wandb.ai/authorize`.

3. Alternatively, you can set the `WANDB_API_KEY` environment variable to your API key, as follows:

    ```
    !export WANDB_API_KEY=e7d*********
    ```

4. Again, keeping the entire code as-is, we only add two parameters, `report_to="wandb"` and `run_name="..."`, to `TrainingArguments`, which enables logging in to W&B, as shown in the following code block:

    ```
    training_args = TrainingArguments(
        ... the rest is same ...
        run_name="IMDB-batch-32-lr-5e-5",
        report_to="wandb"
    )
    ```

5. Then, as soon as you call `trainer.train()`, logging starts on the cloud. After the call, please check the cloud dashboard and see how it changes. Once the `trainer.train()` call has completed successfully, we execute the following line to tell wandB we are done:

```
wandb.finish()
```

The execution also outputs run history locally, as follows:

Run history:

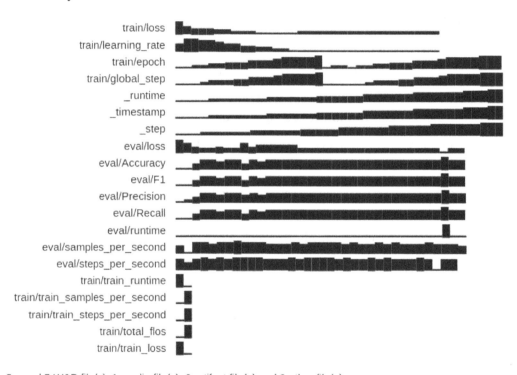

Synced 5 W&B file(s), 1 media file(s), 0 artifact file(s) and 0 other file(s)

Synced .IMyIMDBModel: https://wandb.ai/savasy/huggingface/runs/204bdria

Figure 11.19 – The local output of W&B

When you connect to the link provided by W&B, you will get to an interface that looks something like this:

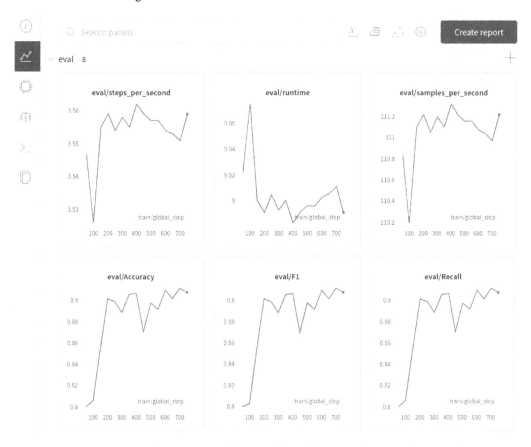

Figure 11.20 – The online visualization of a single run on the W&B dashboard

This visualization gives us a summarized performance result for a single run. As you see, we can trace the metrics that we defined in the `compute_metric()` function.

Now, let's take a look at the evaluation loss. The following screenshot shows exactly the same plot that TensorBoard provided, where the minimum loss is around 0.2658, occurring at step 250:

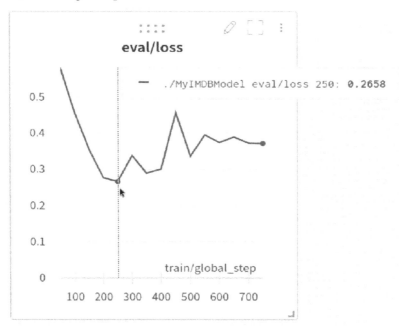

Figure 11.21 – The eval/loss plot of IMDb experiment on the W&B dashboard

We have only visualized a single run so far. W&B allows us to explore the results dynamically across lots of runs at once—for example, we can visualize the results of models using different hyperparameters such as learning rate or batch size. To do so, we instantiate a `TrainingArguments` object properly with another different hyperparameter setting and change `run_name="..."` accordingly for each run.

The following screenshot shows our several IMDb sentiment-analysis runs using different hyperparameters. We can also see the batch size and learning rate that we changed:

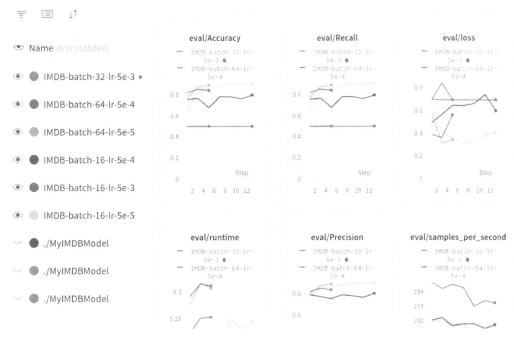

Figure 11.22 – Exploring the results across several runs on the W&B dashboard

W&B provides useful functionality—for instance, it automates hyperparameter optimization and searching the space of possible models, called W&B Sweeps. Other than that, it also provides system logs relating to **Graphics Processing Unit (GPU)** consumption, **Central Processing Unit (CPU)** utilization, and so on. For more detailed information, please check the following website: `https://wandb. ai/home`.

Well done! In the last section, *References*, we will focus more on technical tools, since it's crucial to use such utility tools to develop better models.

Summary

In this chapter, we introduced two different technical concepts: attention visualization and experiment tracking. We visualized attention heads with the exBERT online interface first. Then, we studied BertViz, where we wrote Python code to see three BertViz visualizations: head view, model view, and neuron view. The BertViz interface gave us more control so that we could work with different language models. Moreover, we were also able to observe how attention weights between tokens are computed. These tools provide us with important functions for interpretability and exploitability. We also learned how to track our experiments to obtain higher-quality models and do error analysis. We utilized two tools to monitor training: TensorBoard and W&B. These tools were used to effectively track experiments and to optimize model training.

Congratulations! You've finished reading this book by demonstrating great perseverance and persistence throughout this journey. You can now feel confident as you are well equipped with the tools you need, and you are prepared for developing and implementing advanced NLP applications.

References

- *exBERT: A Visual Analysis Tool to Explore Learned Representations in Transformer Models, Benjamin Hoover, Hendrik Strobelt, Sebastian Gehrmann, 2019.*

- *Vig, J., 2019. A multiscale visualization of attention in the Transformer model. arXiv preprint arXiv:1906.05714.*

- *Clark, K., Khandelwal, U., Levy, O.* and *Manning, C.D., 2019. What does bert look at? An analysis of bert's attention. arXiv preprint arXiv:1906.04341.7*

- *Biewald, L., Experiment tracking with weights and biases, 2020.* Software available from wandb.com, *2(5).*

- *Rogers, A., Kovaleva, O.* and *Rumshisky, A.,2020. A primer in BERTology: What we know about how BERT works. Transactions of the Association for Computational Linguistics, 8, pp.842-866.*

- *W&B:* https://wandb.ai

- *TensorBoard:* https://www.tensorflow.org/tensorboard

- *exBert—Hugging Face:* https://huggingface.co/exbert

- *exBERT:* https://exbert.net/

`Packt.com`

Subscribe to our online digital library for full access to over 7,000 books and videos, as well as industry leading tools to help you plan your personal development and advance your career. For more information, please visit our website.

Why subscribe?

- Spend less time learning and more time coding with practical eBooks and Videos from over 4,000 industry professionals

- Improve your learning with Skill Plans built especially for you

- Get a free eBook or video every month

- Fully searchable for easy access to vital information

- Copy and paste, print, and bookmark content

Did you know that Packt offers eBook versions of every book published, with PDF and ePub files available? You can upgrade to the eBook version at `packt.com` and as a print book customer, you are entitled to a discount on the eBook copy. Get in touch with us at `customercare@packtpub.com` for more details.

At `www.packt.com`, you can also read a collection of free technical articles, sign up for a range of free newsletters, and receive exclusive discounts and offers on Packt books and eBooks.

Other Books You May Enjoy

If you enjoyed this book, you may be interested in these other books by Packt:

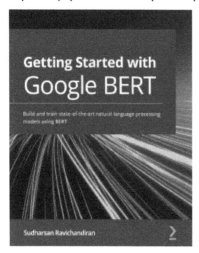

Getting Started with Google BERT

Sudharsan Ravichandiran

ISBN: 978-1-83882-159-3

- Understand the transformer model from the ground up
- Find out how BERT works and pre-train it using masked language model (MLM) and next sentence prediction (NSP) tasks
- Get hands-on with BERT by learning to generate contextual word and sentence embeddings
- Fine-tune BERT for downstream tasks
- Get to grips with ALBERT, RoBERTa, ELECTRA, and SpanBERT models
- Get the hang of the BERT models based on knowledge distillation

Mastering spaCy

Duygu Altınok

ISBN: 978-1-80056-335-3

- Install spaCy, get started easily, and write your first Python script
- Understand core linguistic operations of spaCy
- Discover how to combine rule-based components with spaCy statistical models
- Become well-versed with named entity and keyword extraction
- Build your own ML pipelines using spaCy
- Apply all the knowledge you've gained to design a chatbot using spaCy

Leave a review - let other readers know what you think

Please share your thoughts on this book with others by leaving a review on the site that you bought it from. If you purchased the book from Amazon, please leave us an honest review on this book's Amazon page. This is vital so that other potential readers can see and use your unbiased opinion to make purchasing decisions, we can understand what our customers think about our products, and our authors can see your feedback on the title that they have worked with Packt to create. It will only take a few minutes of your time, but is valuable to other potential customers, our authors, and Packt. Thank you!

Share Your Thoughts

Now you've finished *Mastering Transformers*, we'd love to hear your thoughts! Scan the QR code below to go straight to the Amazon review page for this book and share your feedback or leave a review on the site that you purchased it from.

https://packt.link/r/1-801-07765-7

Your review is important to us and the tech community and will help us make sure we're delivering excellent quality content.

Index

www.ingramcontent.com/pod-product-compliance
Lightning Source LLC
Chambersburg PA
CBHW060923060326
40690CB00041B/3035